Chemical Process Engineering:
Beyond the Basics

Chemical Process Engineering: Beyond the Basics

Edited by Benedict Walsh

CLANRYE
INTERNATIONAL
www.clanryeinternational.com

Clanrye International,
750 Third Avenue, 9th Floor,
New York, NY 10017, USA

ISBN: 978-1-63240-711-5

Cataloging-in-Publication Data

Chemical process engineering : beyond the basics / edited by Benedict Walsh.
    p. cm.
Includes bibliographical references and index.
ISBN 978-1-63240-711-5
1. Chemical processes. 2. Chemical engineering. I. Walsh, Benedict.
TP155.7 .C45 2018
660.28--dc23

For information on all Clanrye International publications
visit our website at www.clanryeinternational.com

CLANRYE
INTERNATIONAL

Contents

Preface

Chemical process engineering deals with the designing and optimization of chemical manufacturing systems. It is concerned with the design and construction of manufacturing systems, which turn raw materials into finished products. Other aspects of chemical process engineering include heat exchanger networks, sustainable industrial activity, etc. Some of the diverse topics covered in this book address the varied branches that fall under this category. As this field is emerging at a rapid pace, the contents of this textbook will help the readers understand the modern concepts and applications of the subject.

A detailed account of the significant topics covered in this book is provided below:

Chapter 1- Process integration is a term used for a comprehensive approach towards different unit operations. Process integration has two main branches, energy integration and mass integration. Mass pinch, water pinch and hydrogen pinch, all parts of pinch analysis, are some of the applications of this field. This is an introductory chapter which will introduce briefly all the significant aspects of chemical process engineering.

Chapter 2- Pinch technology is a technology used for reducing energy consumption in chemical processes. Some of the examples of pinch analysis softwares are pinexo, integration, pinchleni and heatIT. The diverse applications of pinch technology in the current scenario have been thoroughly discussed in this section.

Chapter 3- Pinch design methods are very useful as it allows engineers to incorporate real world problems that occur in industries. Pinch analysis is used for the design of heat exchange networks (HENs). The aspects elucidated in this chapter are of vital importance, and provide a better understanding of chemical process engineering.

Chapter 4- A mathematical model uses concepts of mathematics. These models can be used in various subjects like physics, biology, psychology, sociology and artificial intelligence. The chapter closely examines the key concepts of mathematical modeling to provide an extensive understanding of the subject.

I would like to make a special mention of my publisher who considered me worthy of this opportunity and also supported me throughout the process. I would also like to thank the editing team at the back-end who extended their help whenever required.

Editor

Understanding Process Integration

Process integration is a term used for a comprehensive approach towards different unit operations. Process integration has two main branches, energy integration and mass integration. Mass pinch, water pinch and hydrogen pinch, all parts of pinch analysis, are some of the applications of this field. This is an introductory chapter which will introduce briefly all the significant aspects of chemical process engineering.

Process Integration

Often, underlying principles behind technical innovations can be found in the realm of nature. This is also true for process integration as well, if its meaning in a broad spectrum is considered. Following examples make it amply clear.

Many of you must have observed that many birds fly in flocks. Geese fly in a variety of formations, ranging from slanted lines topatterns that look like V's, U's, or even W's. They have a tendency to fly close to each other, in the same horizontal plane. Have you ever thought why they fly like this? There are a number of social factors such as reproduction, communication, navigation, protection from predators, etc. However, if only aerodynamic factors are considered for the time being, one apparent theory is that in coordinated flight flocks such as waterfowl, there's an aerodynamic advantage to flying behind and tothe side of another bird to take advantage of its wingtip vortices. The next time you see ducks or geese flying in a V- formation, you will observe that the lead bird changes its position periodically. This is for the reason that who so ever is on the front takes the maximum drag resistanceof the wind. That is why the leader takes a back position after a while, where wind drag is the lowest. The bird who has taken sufficient rest comes to the front to face the hardship of the flight. Birds also fly in other formations as a strategy for dealing with wind also.

Thus, birds fly in a V-formation (in an integrated form) to help conserve total energy of the flock during migrations. This is an excellent example of flight integration where every bird is benefited. Further, this has been noticed that birds only migrate when there are favorable wind conditions (it takes 40% less energy for birds riding on up drafts than the energy needed for lift if they are flying on their own). In short, to save energy, the winds must be blowing in the direction that the birds want to fly. Birds will usually wait until the most favorable weather conditions then only set off on their journey. This clearly shows that when birds integrate their flying with surrounding winds they are further benefited.

Another example of integration in nature is that fish swim in schools. Schools are composed of many fish of the same species moving in more or less harmonious patterns throughout the oceans. A very widespread behavior, schooling, is exhibited by almost 80 percent of the more than 20,000 known fish species during some phase of their life cycle. Aristotle, over 2,400 years ago, observed this behavior in fish

But why fish swim in schools? Besides a large number of social factors such as safety against predators, there are advantages in terms of energy consumption too. Some species of fish secrete a "slime" that helps in reducing the friction of water over their bodies. Also fishes swim in fairly precise, staggered patterns when traveling in schools and the "to-and-fro" motion of their tails producestiny currents called "vortices" (swirling motions similar to little whirlpools). Each individual, in theory, can use the tiny whirlpool of its neighbor to assist in reducing the water's friction on its own body.

Another glaring example of process integration in our society is the concept of Joint Family. It certainly decreases the per member expenditure. Other important example of process integration is "Sanjha Chulha" or common burner which had been a tradition of Punjab where many family cook on the same to reduces the use of fossil fuels. Yet another example is cooking of meals in the famous "Jagannath" temple of Puri, Odisha. Jagannath Temple at Puri in India is said to have the largest kitchen in the world. The kitchen prepares food for 100,000 people on a festival day and for about 25,000 on a normal day. The unique feature of cooking in this temple is that, clay pots are placed in a special earthen oven, five in numbers, one on the top of another there by saving a lot of energy in cooking. A similar principle used in multi-effect evaporation system. The above tradition is prevalent since 11th century.

Even the concept of Integration is visible in microorganisms. This is known as symbiosis. It is a relationship between individuals of different species where both individuals derive a benefit from living together. One of the most spectacular example of symbiosis is between "siboglinidtube worms" and "symbiotic bacteria" that live at hydrothermal vents and cold seeps. The worm has no digestive tract and is wholly reliant on its internal symbionts for nutrition. The bacteria oxidize either hydrogen sulfide or methane which the host supplies to them.

A considerable number of examples can be found in nature which shows that an integrated process takes less energy to perform an assigned task. But the question is how to use this concept in process industry?

The examples given below use the concepts of process integration in process industry examples. One such example of integration in process industry can be seen in the sixteenth century technology of gold retrieval as described below :

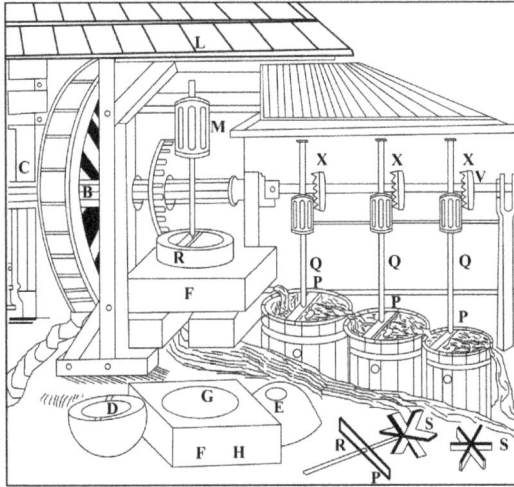

The technology is very simple. The ore is crushed by the stamp, "C," ground in the mill, "F," andmixed with mercury in vessels "O." Gold is extracted from the ore by mercury andis later separated from it by pressing the mixture through a leather or cloth filterbag (not shown in the drawing).

Taking a closer look at the woodcut, one noticesthat the stamp, the mill and the stirrers for mixing the ore with mercury are alldriven by the same water wheel, "A," via the common axle, "B," and a number ofvarious gears. In the language of the 21st century, one could say, "A marvelous example of a green, energy-based, highly integrated processing plant.

As a second example let us consider a front end specialty chemical process illustrated in the Figure below :

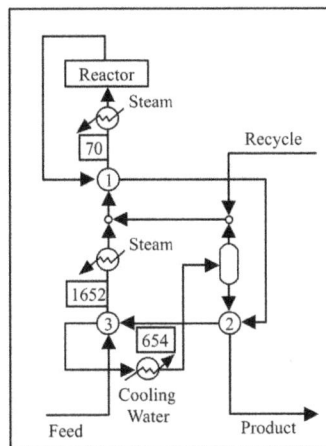

The process uses six heat transfer units (2 heaters, 1 cooler and 3 heat exchanger units and has an energy requirement of 1722 kW of heating and 654 kW of cooling. From the first appearance it appears to be thermally efficient whereas it is not so. So, what is wrong with it?

The answer is that it lacks effective integration. This will become clear from the figure shown below:

This configuration uses only four heat transfer units and the utility heating load is reduced by around 40 % (1068 kW) with cooling no longer required. The design is safe and operable as the earlier one. It is simply better, as it has most effective energy integration.

Another inspiringexample of process integration is shown in Figure. The figure shows a coal pyrolysis process. The main products of which are different hydrocarbon cuts. Benzene obtained from pyrolysis is further processed in a hydrogenation reactor to produce cyclohexane. A hydrogen-rich gas is produced from the cyclohexane reactor and is currently flared.

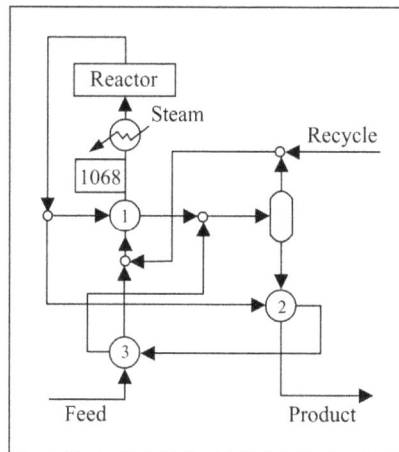

Further the process demands that, medium and heavy distillates contain objectionable materials (primarily sulfur, but also nitrogen, oxygen, halides) should be removed and unsaturated hydrocarbons (e.g. olefins and gum-forming unstable di-olefins) should be converted to paraffins.

One way of addressing this problem is to establish hydro treating and hydro desulfurization units requiring fresh hydrogen to remove the objectionable material and to stabilize olefins and di-olefins as shown in the Figure.

But, the drawback of this process is that there is no integration of mass. At one side, fresh hydrogen is being purchased and used in hydro treating and desulfurization and on the other side; hydrogen produced from benzene dehydrogenation is flared. Therefore, to improve the economics of the process, process integration is needed in this case to conserve the resources.

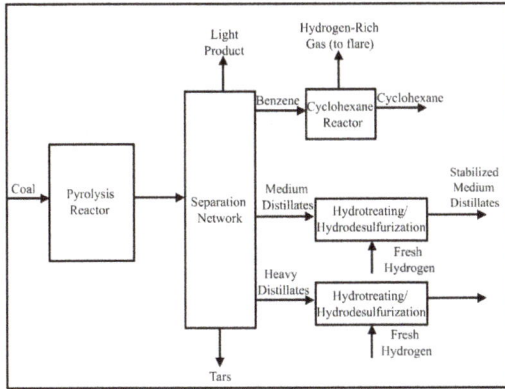

In addition to integration with in a process, we often come across examples where significant benefits have been obtained from integration between different processes.One such example of multi-site integration where sites operating different processes are linked indirectly through utilities is shown below in Figure: In this scheme exhaust steam of power plant is used in Processes A, B and C

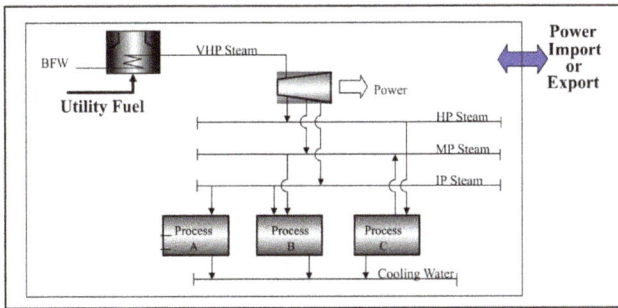

Another example of multi process integration is the integration between steel plant and oil refining plant. In this case, coke oven gas is fed to hydrogen generation unit to increase its production. In this way, the otherwise waste gas is used for producing useful chemicals like gasoline and gas oil. The detailed process is shown in Figure below

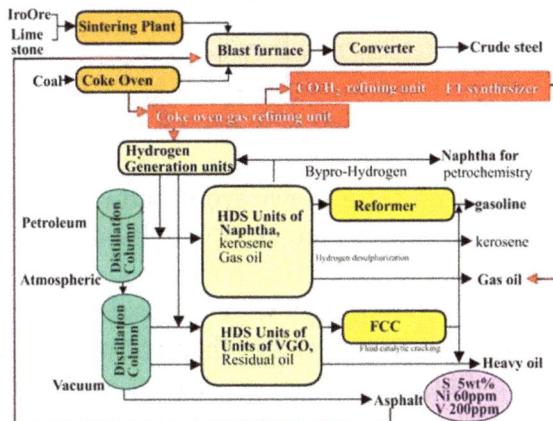

After establishing the fact that process integration will certainly help in reducing the utility and material consumption of process industry, now let us see what the extent of energy loss from different industries is? Figure shows the extent of energy loss from different industries which clearly offers necessary impetus to save energy losses.

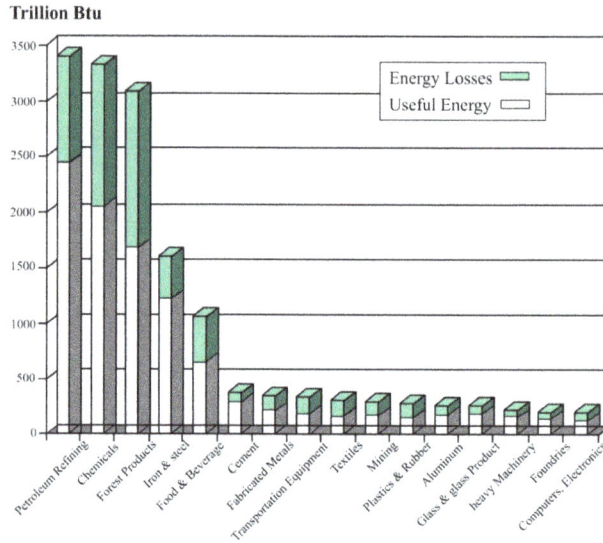

After observing clearly the benefits of process integration in process industries, let us examine potential benefits which can be obtained from the application of process integration in different industriesas reported in figure obelow:

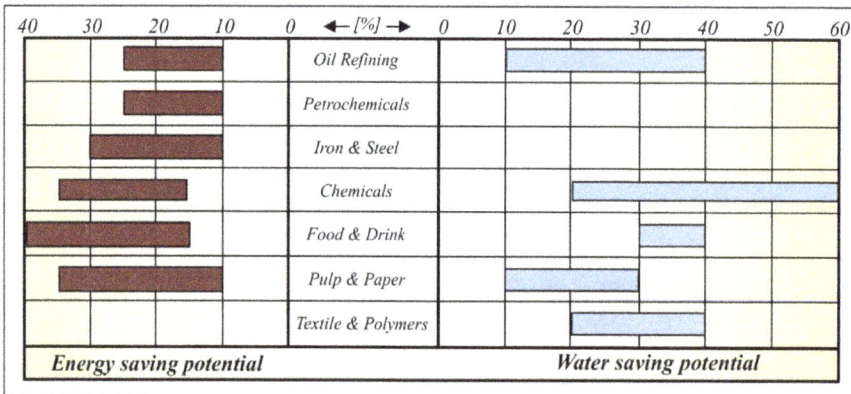

From the above examples, which were drawn from nature as well as existing industries, it is now amply clear that process integration helps industry to conserve energy and material. Conceptually Process integration is a term in chemical engineering which means a holistic approach to process design which considers the interactions between different unit operations from the outset, rather than optimizing them separately. This can also be called integrated process design or process synthesis. Smith (2005) has described the approach well.

Now the question is whether there is a systematic analysis technique which will guide us step by step to apply process integration principles in industries. The answer is yes.

Process integration is a term in chemical engineering which has two possible meanings.

1. A holistic approach to process design which emphasizes the unity of the process and considers the interactions between different unit operations from the outset, rather than optimising them separately. This can also be called *integrated process design* or *process synthesis*. El-Halwagi (1997 and 2006) and Smith (2005) describe the approach well. An important first step is often *product design* (Cussler and Moggridge 2003) which develops the specification for the product to fulfil its required purpose.

2. *Pinch analysis*, a technique for designing a process to minimise energy consumption and maximise heat recovery, also known as *heat integration, energy integration* or *pinch technology*. The technique calculates thermodynamically attainable *energy targets* for a given process and identifies how to achieve them. A key insight is the pinch temperature, which is the most constrained point in the process. The most detailed explanation of the techniques is by Linnhoff et al. (1982), Shenoy (1995) and Kemp (2006). This definition reflects the fact that the first major success for process integration was the thermal pinch analysis addressing energy problems and pioneered by Linnhoff and co-workers. Later, other pinch analyses were developed for several applications such as mass-exchange networks (El-Halwagi and Manousiouthakis, 1989), water minimization (Wang and Smith, 1994), and material recycle (El-Halwagi et al., 2003). A very successful extension was "Hydrogen Pinch", which was applied to refinery hydrogen management (Nick Hallale et al., 2002 and 2003). This allowed refiners to minimise the capital and operating costs of hydrogen supply to meet ever stricter environmental regulations and also increase hydrotreater yields.

In the context of chemical engineering, Process Integration can be defined as a holistic approach to process design and optimization, which exploits the interactions between different units in order to employ resources effectively and minimize costs.

Note that Process Integration is not limited to the design of new plants, but it also covers retrofit design (e.g. new units to be installed in an old plant) and the operation of existing systems. Nick Hallale (2001), in his article in Chemical Engineering Progress provided a state of the art review. He explained that process integration far wider scope and touches every area of process design. Industries are making more money from their raw materials and capital assets while becoming cleaner and more sustainable.

The main advantage of process integration is to consider a system as a whole (i.e. integrated or holistic approach) in order to improve their design and/or operation. In contrast, an analytical approach would attempt to improve or optimize process units separately without necessarily taking advantage of potential interactions among them.

For instance, by using process integration techniques it might be possible to identify that a process can use the heat rejected by another unit and reduce the overall energy consumption, even if the units are not running at optimum conditions on their own. Such an opportunity would be missed with an analytical approach, as it would seek to optimize each unit, and there after it wouldn't be possible to re-use the heat internally.

Typically, process integration techniques are employed at the beginning of a project (e.g. a new plant or the improvement of an existing one) to screen out promising options to optimize the design and/or operation of a process plant.

Also it is often employed, in conjunction with simulation and mathematical optimization tools to identify opportunities in order to better integrate a system (new or existing) and reduce capital and/or operating costs.

Most process integration techniques employ Pinch analysis or Pinch Tools to evaluate several processes as a whole system. Therefore, strictly speaking, both concepts are not the same, even if in certain contexts they are used interchangeably. The review by Nick Hallale (2001) explains that in the future, several trends are to be expected in the field. In the future, it seems probable that the boundary between targets and design will be blurred and that these will be based on more structural information regarding the process network. Second, it is likely that we will see a much wider range of applications of process integration. There is still much work to be carried out in the area of separation, not only in complex distillation systems, but also in mixed types of separation systems. This includes processes involving solids, such as flotation and crystallization. The use of process integration techniques for reactor design has seen rapid progress, but is still in its early stages. Third, a new generation of software tools is expected. The emergence of commercial software for process integration is fundamental to its wider application in process design.

Methods and Application

Process integration (PI), a part of Process Intensification, is a fairly new term that emerged in 80's and has been extensively used in the 90's to describe certain systems oriented activities related primarily to cover almost complete process design. Process integration is a holistic approach to process design, retrofitting, and operation of industrial plants, with applications focused on resource conservation, pollution prevention and energy management. Two main branches of process integration can be recognized as: Energy integration, that deals with the global allocation, generation, and exchange of energy throughout the process and Mass integration that provides a fundamental understanding of the global flow of mass within the process and optimizes the allocation, separation, and generation of streams and species.

PI is a smart framework for the holistic analysis of process performance and the generation of cost-effective and sustainable solution strategies. It is based on fundamental chemical engineering and systems principles and therefore provides a set of generally-applicable tools. It enables the process engineer to see "the big picture first, and the details later". With this approach, it is not only possible to identify the optimal process development strategy for a given task but also to uniquely identify the most cost- effective way to accomplish that task. Generally speaking, PI is concerned to the advanced management of material, energy and information flows in a production plant and the surrounding community based on the multi criteria optimization of the processing systems.

Process integration has already had a profound effect on the chemical process industries, in the form of pinch technology and heat-exchanger-network optimization. However, it has mistakenly been interpreted as Heat Integration by many people, probably due to the fact that heat recovery studies inspired by Pinch Concept initiated this field and still remains the central part of Process Integration. Process integration appears to be a rather dynamic field, with new method and application areas emerging constantly into new areas, like Mass Pinch, Water Pinch, Hydrogen pinch, by using various analogies. Another interesting example of process integration is a dividing-wall column, which essentially integrates two distillation columns into one, thereby eliminating two pieces of capital equipment — the condenser from the first column and the reboiler from the second one.

Process Integration Definitions

Over the years there have been several attempts to define process integration. A study of the most well-known definitions reveals that it has become difficult to describe the fundamental principle behind process integration :

- In 1993 the International Energy Agency (IEA) defined process integration as: Systematic and general methods for designing integrated production systems, ranging from individual processes to total sites, with special emphasis on the efficient use of energy and reducing environmental effects (Gundersen, 2002). By this definition, process integration is seen as a group of methods to optimize the use of energy, but with concerns for environmental aspects.

- In 1997 the IEA broadened their definition of process integration to mean the application of methodologies developed for system-oriented and integrated approaches to industrial process plant design for both new and retrofit applications (Gundersen, 1997). Along with this the optimization of the system became a goal and a need for the method's applicability throughout the life cycle was recognized.

- Later, Natural Resources Canada (2003) defined process integration as all improvements made to process systems, their constituent unit operations, and their interactions to maximize the effective use of energy, water and raw materials.

- In the Finnish process integration technology program, process integration was defined to mean: Integrated and system-oriented planning, operation and the optimization and management of industrial processes. The operation and management aspects are emphasized in the Finnish definition. The above definitions describe the objectivity of a process integration task rather than the principles through which the enhanced situation is achieved.

- Rossiter and Kumana (1995) state that process integration methods includes, focus on ensuring that existing process technologies are selected and interconnected in the most effective ways rather than attempting to invent new types of equipment or unit operations. This definition slightly touches the potential synergic effects which will be achieved by integration.

- According to the definition by El-Halwagi (1997), integration emphasizes the unity of the process. According to him, "Process integration is a holistic approach to process design, retrofitting, and operation which emphasizes the unity of the process."

- According to Ferenc Friedler, 2010, Process integration is a family of methodologies for combining several processes to reduce consumption of resources or harmful emissions to the environment.

Many definitions of PI can be found in the literature, but the most complete is the one used by the IEA since 1993 which states that "systematic and general methods for designing integrated production systems, ranging from individual processes to total sites, with special emphasis on the efficient use of energy and reducing environmental effects".

This definition brings Process Integration very close to Process Synthesis, which is another systems oriented technology. Process Integration and synthesis belongs to process systems engineering.

From History to the Future

Process Design has evolved through distinct "generations". Originally (first generation), inventions that were based on experiments in the laboratory by the chemists, which were then tested in pilot plants before plant construction.

The second generation of Process Design was based on the concept of Unit Operations, which founded Chemical Engineering as a discipline. Unit Operations acted as building blocks for the engineer in the design process.

The third generation considered integration between these units; for example heat recovery between related processes streams to save energy.

A strong trend today (fourth generation) is to move away from Unit Operations and focus on Phenomena. Processes based on the Unit Operations concept tend to have many process units with significant and complex piping arrangements between the units. By allowing more than one phenomena (reaction, heat transfer, mass transfer, etc.) to take place within the same piece of equipment, significant savings have been observed both in investment cost and in operating cost (energy and raw materials). However, most of the industrial applications of this idea have been based on trial and error. In this area the research is making progress, with an aim to develop systematic methods to replace trial and error. No doubt, this will affect the discipline of Process Integration, since we no longer look at integration between units only, but also at integration within units and within a site.

Different Schools of Thoughts in Process Integration

The three major features of Process Integration methods are the use heuristics (insight), about design and economy, the use of thermodynamics and the use of optimization techniques. There is significant overlap between the various methods and the trend today is strongly towards methods using all three features mentioned above. The large number of structural alternatives in Process Design (and Integration) is significantly reduced by the use of insight, heuristics and thermodynamics, and it then becomes feasible to address the remaining problem and its multiple economic trade- offs with optimization techniques.

Despite the merging trend mentioned above, it is still valid to say that Pinch Analysis and Exergy Analysis are methods with a particular focus on Thermodynamics. Hierarchical Analysis and Knowledge Based Systems are rule-based approaches with the ability to handle qualitative (or fuzzy) knowledge. Finally, Optimization techniques can be divided into deterministic (Mathematical Programming) and non- deterministic methods (stochastic search methods such as Simulated Annealing and Genetic Algorithms). One possible classification of Process Integration methods is to use the two-dimensional (automatic vs. interactive and quantitative vs. qualitative) representation in Figure.

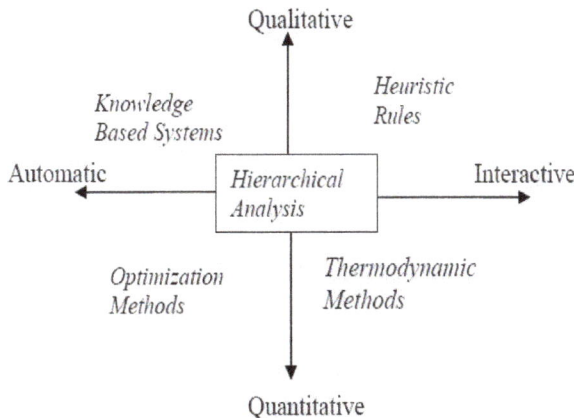

One possible Classification of Process Integration

Application of Process Integration

Process Integration concepts can be applied in various fields such as:

1. Heat integration – heat exchange network

2. Distillation column targeting

3. Cogeneration and total site targeting

4. Batch process targeting and optimization

5. Emission targeting(GHG emission reduction)

6. Mass exchange network (water and waste water management & recovery of valuable materials)

7. Hydrogen management in refineries

8. Debottlenecking of critical areas in process industries.

9. Pollution prevention

10. E- Education system

11. Co-production system

12. Low temperature process

13. supply-chain management

14. Financial management

15. Carbon-constrained energy-sector planning

Techniques Available for Process Integration

1   Pinch Technology Approach

2. MILP/MINLP Approach

3. State-Space Approach

4. Genetic Algorithm Approach

5. Process Graph Theory Approach

Current Status of Process Integration

Process Integration is a strongly growing field of Process Engineering. It is now a standard curriculum for process engineers in both Chemical and Mechanical Engineering at most universities around the world, either as a separate topic or as part of a Process Design or Synthesis course. Research at UMIST in this area has for last 27 years been supported by a large number of industrial companies through a Consortium that was established in 1984. As part of the International Energy Agency (IEA) project on Process Integration, 35 other universities around the world are involved in this research field.

Process Integration has evolved from a Heat Recovery methodology in the 80's to become what a number of leading industrial companies in the 90's regarded as a Major Strategic Design and Planning Technology. Process integration, combined with other tools such as process simulation, is a powerful approach that allows engineers to systematically analyze an industrial process and the interactions between its various parts. Process integration design tools have been developed over the past two decades to achieve process improvement, productivity enhancement, conservation in mass and energy resources, and reductions in the operating and capital costs of chemical processes. The primary applications of these integrated tools have focused on resource conservation, pollution prevention and energy management.

Specifically, the past two decades have seen the development and/or application of process integration design tools for heat exchange networks (HENs), wastewater reduction and water conservation networks, mass exchange networks (MENs), heat- and energy-induced separation networks (HISENs and EISENs), waste interception networks (WINs) and heat-and energy-induced waste minimization networks (HIWAMINs and EIWAMINs), Hydrogen management to name a few.

Pinch analysis techniques have also been extended to various carbon and environmental constrained problems. The first applications were meant to determine the minimum amount of zero or low carbon energy sources needed to meet the regional or sectorial emission limits. The concept was later extended to segregate targeting with regions using unique sets of energy sources, and for targeting retrofits for carbon sequestration in the electricity sector. Furthermore, the pinch analogy was used for energy planning in scenarios involving land and water footprints.

Pinch Technology

Pinch Analysis is a methodology for minimizing energy consumption of processes by calculating thermodynamically feasible energy targets (or minimum energy consumption) and achieving them by optimizing heat recovery systems, energy supply methods and process operating conditions. It is also known as "process integration", "heat integration", "energy integration" or "pinch technology".

Process integration has already had a profound effect on the chemical process industries, in the form of Pinch Technology and heat-exchanger-network optimization. "Pinch analysis (for energy) has had an enormous amount of application, with thousands of projects having been carried out all over the world. AspenTech's Nick Hallale, has reported that "Pinch analysis [for energy] has had an enormous amount of application, with thousands of projects having been carried out all over the world. Companies, such as Shell,Exxon, BP-Amoco, Neste Oy, and Mitsubishi, have reported fuel savings of upto 25% and similar emissions reductions, worth millions of dollars per year."

Among the PI methodologies, Pinch Analysis is currently the most widely used. This is due to the simplicity of its underlying concepts and, specially, to the spectacular results it has obtained in numerous projects worldwide.

Pinch technology is a rigorous, structured thermodynamic approach to energy efficiency that can be used to tackle a wide range of process and utility related problems, such as reducing operating costs, debottlenecking processes, improving efficiency and reducing and planning capital investment.

The term "Pinch Technology" was introduced by Linnhoff and Vredeveld in 1979 to represent a new set of thermodynamically based methods that guarantee minimum energy levels in the design of heat exchanger networks. It is a systematic methodology based on thermodynamic principles to achieve utility savings by better process heat integration, maximizing heat recovery and reducing the external utility loads (cooling water and heating steam). Over the last two decades it has emerged as an unconventional development in process design and energy conservation. The term 'Pinch Technology' is often used to represent the application of the tools and algorithms of Pinch analysis for studying industrial processes. Pinch Technology is a recognized and well proven method in industries such as chemical, petrochemical, oil refining, paper and pulp, food and drinks, steel and metallurgy, etc., leading to an energy saving of 10 to 35%, water saving of the tune of 25 to 40% and hydrogen savings up to 20%. Pinch technology provides a systematic methodology for energy saving in processes and total sites.

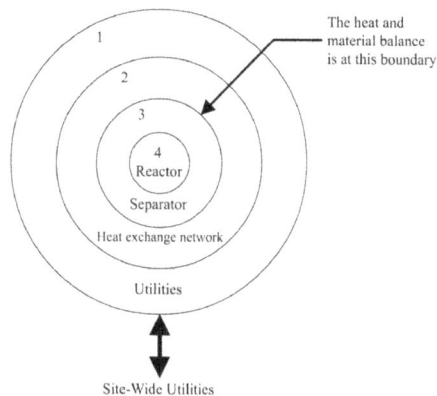

1

2

3

4
Reactor

Separator

Heat exchange network

Utilities

The heat and material balance is at this boundary

Site-Wide Utilities

Energy saving of 10 to 35%, water saving of the tune of 25 to 40% and hydrogen savings up to 20%. Pinch technology provides a systematic methodology for energy saving in processes and total sites

Figure illustrates the role of Pinch Technology in the overall process design. The process design hierarchy can be represented by the "onion diagram " as shown below. The design of a process starts with the reactors (in the "core" of the onion which brings chemical changes in the feed). Once feeds, products, recycle concentrations and flow rates are known, the separators (the second layer of the onion) can be designed and based on the requirement of heat for the core and second layer the heat exchange network (the third layer) can be designed. The remaining heating and cooling duties are handled by the utility system (the fourth layer). The process utility system may be a part of a centralized site-wide utility system. The connectivity of the onion diagram with actual process is demonstrated in Figure.

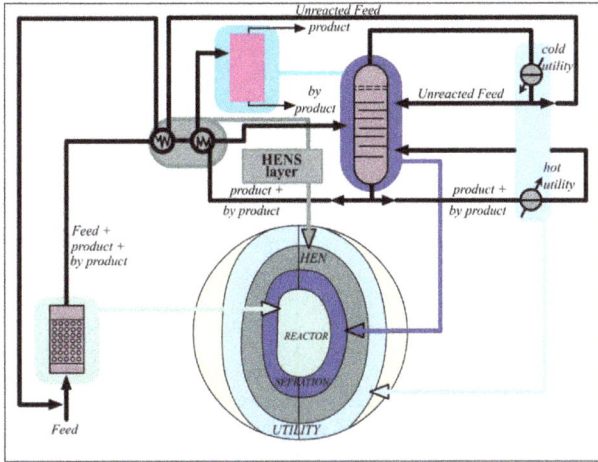

A Pinch Analysis starts with the heat and material balance for the process. Using Pinch Technology, it is possible to identify appropriate changes in the core process conditions that can have an impact on energy savings (onion layers one and two). After the heat and material balance is established, targets for energy saving can be set prior to the design of the heat exchanger network.

Pinch analysis deals with about sixty principles and concepts as given in table.

Table: Principles tools and design rules of Pinch Analysis.

Principles and Tools	Rules
Basic HEN Design	
Composite Curves Problem Table Threshold Plot Bath Area Target Delta −T Contribution Euler & No. of Units Target Grid Diagram Pinch Principle CP Matrix Criss Cross Principle Driving Force Plot(s) Remaining Problem Analysis Loops & Paths Supertargeting Delta-p Targeting and Supertargeting Retrofit Targeting (constant α) Area Matrix	Pinch Design Method Stream Splitting Rules Mixing Rules HEN Evolution Topology Traps Retrofit Design

Utility Targeting and Design	
Grand Composite Curve	Furnace Integration
Utility Composite & Utility Pinches Balanced Composites & Balanced Grid Total Site Profiles (T-H)	Heat Engine Placement Heat Pump Placement Multiple Utility Optimization Low T Process Design
Cooling Water Targets	
Exergy & Pinch : The Nc-H Plot	
Exergy Grand Composites Low T Shaftwork Target Total Site Profiles (Nc-H) Power Cycle Targeting	
Advanced HEN Design	
Rigorous Area Targets	Constrained HEN Design Condensing Steam Cycle Power Block
Constrained HEN Targeting	Design
Area Matrix Retrofit Targets	Diverse pinch concept for heat exchange network synthesis
Area Integrity Matrix	
Area Cost Targets for Different Mils for Construction	
No. of Shells Targets Downstream Paths Sensitivity Tables Multiple Base Case Targets	
Re-piping & Rerouting Targets Non-Convexity	
Resiliency Index Time Slice Targets Cascade Analysis Batch Utility Curves	

Process Design	
The Onion Model Keep Hot Streams Hot & Cold Streams Cold	Data Extraction Rules Appropriate Placement of Distillation Columns Appropriate Placement of Evaporators
Plus/Minus Principle Column Grand Composite Curves Column Composite Curves	Waste Water System Design Rules
Waste Water Targets	

Now this methodology has become broad based while maintaining its principles still those of heat & power and thermodynamics and its key strategy is to set targets prior to design. Now it addresses systems including distillation, heat pumps, co-generating turbines, furnaces and non-energy objectives such as capital cost, operability and emissions. Hallale has developed the newest member of the pinch family — hydrogen pinch — which is aimed at helping oil refiners better manage their hydrogen balances. Although new, the technology has already had several applications and resulting millions of dollars of savings per year.

Uday V. Shenoy (2011) suggested a single algorithm to establish minimum resource targets for diverse process integration problems including those of heat/mass exchange, water, hydrogen, carbon emission and material reuse networks and proposed unified targeting algorithm (UTA). His analogies for diverse process integration problems are given in Table.

Table: Analogies for application of unified targeting algorithm (UTA) to diverse integration problems

(a) Heat and Mass integration		
Variables	Heat Exchange Network	Mass Exchange Network
Level or Quality, unit	Temperature, °C	Composition as mass ratio
Flow	Heat capacity flow rate CP, kW/°C	Mass Flow rate, kg/s
Load or quantity	Heat Load, kW	Mass load, kg/s
High Level Resource/Utility	Hot utility	Process MSAs excess
Low Level Resource/Utility	Cold Utility	External MSAs
Level shifting	$\Delta T_{min}/2$	$\Delta y_{min}/2$ or $\varepsilon/2$
Level sort order(preferred)	Decreasing	Decreasing

(b) Water, hydrogen and carbon emission networks			
	Water Networks	Hydrogen Network	Carbon emission Network
Level or quality	Contaminants	Hydrogen purity	Emission factor
Unit	Concentration in ppm	ratio	tCO_2/TJ
Flow	Water flow rate t/h	H_2 flow rate MMscfd or mol/s	Energy in TJ
Load or quantity,	Contaminant load kg/h	H_2 load MMscfd or mol/s	Emission load or carbon foot print, tCO_2
High quality resource	Fresh Water	Make-up H_2 utility	Zero carbon or low carbon resource
Low quality resource	Waste water	Hydrogen purge	Unused or excess energy
Level shifting	0	0	0
Level sort order preferred	Increasing	Increasing	Increasing

Sharifah R. Wan Alwi and Zainuddin A. Manan(2010) presented STEP (Stream Temperature vs. Enthalpy Plot) as a new graphical tool for simultaneous targeting and design of a HEN that overcomes the key limitations of Composite Curves and the Grid Diagram. The new STEPs are profiles of continuous individual hot and cold streams being mapped on a shifted temperature versus enthalpy diagram that simultaneously show the pinch points, energy targets and the maximum heat allocation (MHA). There work also demonstrates that STEP can provide more realistic solutions for targeting multiple utilities and the minimum network area.

It is ample clear that process integration technique can result in huge benefits in process industries and therefore, significant amount of research work is going on throughout the world both in academic as well as commercial arena on process integration.

Process Design

In chemical engineering, process design is the design of processes for desired physical and/or chemical transformation of materials. Process design is central to chemical engineering, and it can be considered to be the summit of that field, bringing together all of the field's components.

Process design can be the design of new facilities or it can be the modification or expansion of existing facilities. The design starts at a conceptual level and ultimately ends in the form of fabrication and construction plans.

Process design is distinct from equipment design, which is closer in spirit to the design of unit operations. Processes often include many unit operations.

Documentation

Process design documents serve to define the design and they ensure that the design components fit together. They are useful in communicating ideas and plans to other engineers involved with the design, to external regulatory agencies, to equipment vendors and to construction contractors.

In order of increasing detail, process design documents include:

- Block flow diagrams (BFD): Very simple diagrams composed of rectangles and lines indicating major material or energy flows.

- Process flow diagrams (PFD): Typically more complex diagrams of major unit operations as well as flow lines. They usually include a material balance, and sometimes an energy balance, showing typical or design flowrates, stream compositions, and stream and equipment pressures and temperatures.

- Piping and instrumentation diagrams (P&ID): Diagrams showing each and every pipeline with piping class (carbon steel or stainless steel) and pipe size (diameter). They also show valving along with instrument locations and process control schemes.

- Specifications: Written design requirements of all major equipment items.

Process designers also typically write operating manuals on how to start-up, operate and shut-down the process.

Documents are maintained after construction of the process facility for the operating personnel to refer to. The documents also are useful when modifications to the facility are planned.

A primary method of developing the process documents is process flowsheeting.

Design Considerations

There are several considerations that need to be made when designing any chemical process unit. Design conceptualization and considerations can begin once product purities, yields, and throughput rates are all defined.

Objectives that a design may strive to include:

- Throughput rate

- Process yield

- Product purity

Constraints include:

- Capital cost

- Available space

- Safety concerns

- Environmental impact and projected effluents and emissions

- Waste production

- Operating and maintenance costs

Other factors that designers may include are:

- Reliability

- Redundancy

- Flexibility

- Anticipated variability in feedstock and allowable variability in product.

Sources of Design Information

Designers usually do not start from scratch, especially for complex projects. Often the engineers have pilot plant data available or data from full-scale operating facilities. Other sources of information include proprietary design criteria provided by process licensors, published scientific data, laboratory experiments, and input.

Computer Help

The advent of low cost powerful computers has aided complex mathematical simulation of processes, and simulation software is often used by design engineers. Simulations can identify weaknesses in designs and allow engineers to choose better alternatives.

However, engineers still rely on heuristics, intuition, and experience when designing a process. Human creativity is an element in complex designs.

Unit Operation

In chemical engineering and related fields, a unit operation is a basic step in a process. Unit operations involve a physical change or chemical transformation such as separation, crystallization, evaporation, filtration, polymerization, isomerization, and other reactions. For example, in milk processing, homogenization, pasteurization, chilling, and packaging are each unit operations which are connected to create the overall process. A process may require many unit operations to obtain the desired product from the starting materials, or feedstocks.

An ore extraction process broken into its constituent unit operations
(Quincy Mine, Hancock, MI ca. 1900)

History

Historically, the different chemical industries were regarded as different industrial processes and with different principles. Arthur Dehon Little propounded the concept of "unit operations" to explain industrial chemistry processes in 1916. In 1923, William H. Walker, Warren K. Lewis and William H. McAdams wrote the book *The Principles of Chemical Engineering* and explained that the variety of chemical industries have processes which follow the same physical laws. They summed up these similar processes into unit operations. Each unit operation follows the same physical laws and may be used in all relevant chemical industries. For instance, the same engineering is required to design a mixer for either napalm or porridge, even if the use, market or manufacturers are very different. The unit operations form the fundamental principles of chemical engineering.

Chemical Engineering

Chemical engineering unit operations consist of five classes:

1. Fluid flow processes, including fluids transportation, filtration, and solids fluidization.

2. Heat transfer processes, including evaporation and heat exchange.

3. Mass transfer processes, including gas absorption, distillation, extraction, adsorption, and drying.

4. Thermodynamic processes, including gas liquefaction, and refrigeration.

5. Mechanical processes, including solids transportation, crushing and pulverization, and screening and sieving.

Chemical engineering unit operations also fall in the following categories which involve elements from more than one class:

- Combination (mixing)

- Separation (distillation, crystallization)

- Reaction (chemical reaction)

Furthermore, there are some unit operations which combine even these categories, such as reactive distillation and stirred tank reactors. A "pure" unit operation is a physical transport process, while a mixed chemical/physical process requires modeling both the physical transport, such as diffusion, *and* the chemical reaction. This is usually necessary for designing catalytic reactions, and is considered a separate discipline, termed chemical reaction engineering.

Chemical engineering unit operations and chemical engineering unit processing form the main principles of all kinds of chemical industries and are the foundation of designs of chemical plants, factories, and equipment used.

In general, unit operations are designed by writing down the balances for the transported quantity for each elementary component (which may be infinitesimal) in the form of equations, and solving the equations for the design parameters, then selecting an optimal solution out of the several possible and then designing the physical equipment. For instance, distillation in a plate column is analyzed by writing down the mass balances for each plate, wherein the known vapor-liquid equilibrium and efficiency, drip out and drip in comprise the total mass flows, with a sub-flow for each component.

Combining a stack of these gives the system of equations for the whole column. There is a range of solutions, because a higher reflux ratio enables fewer plates, and vice versa. The engineer must then find the optimal solution with respect to acceptable volume holdup, column height and cost of construction.

An Integrated Study of Pinch Technology

Pinch technology is a technology used for reducing energy consumption in chemical processes. Some of the examples of pinch analysis softwares are pinexo, integration, pinchleni and heatIT. The diverse applications of pinch technology in the current scenario have been thoroughly discussed in this section.

Pinch Analysis

Pinch analysis is a methodology for minimising energy consumption of chemical processes by calculating thermodynamically feasible *energy targets* (or minimum energy consumption) and achieving them by optimising heat recovery systems, energy supply methods and process operating conditions. It is also known as *process integration*, *heat integration*, *energy integration* or *pinch technology*.

The process data is represented as a set of energy flows, or streams, as a function of heat load (kW) against temperature (deg C). These data are combined for all the streams in the plant to give *composite curves*, one for all *hot streams* (releasing heat) and one for all *cold streams* (requiring heat). The point of closest approach between the hot and cold composite curves is the *pinch point* (or just *pinch*) with a hot stream pinch temperature and a cold stream pinch temperature. This is where the design is most constrained. Hence, by finding this point and starting the design there, the energy targets can be achieved using heat exchangers to recover heat between hot and cold streams in two separate systems, one for temperatures above pinch temperatures and one for temperatures below pinch temperatures. In practice, during the pinch analysis of an existing design, often cross-pinch exchanges of heat are found between a hot stream with its temperature above the pinch and a cold stream below the pinch. Removal of those exchangers by alternative matching makes the process reach its *energy target*.

History

In 1971, Ed Hohmann stated in his PhD that 'one can compute the least amount of hot and cold utilities required for a process without knowing the heat exchanger network that could accomplish it. One also can estimate the heat exchange area required'.

In late 1977, Ph.D. student Bodo Linnhoff under the supervision of Dr John Flower at the University of Leeds showed the existence in many processes of a heat integration bottleneck, 'the pinch', which laid the basis for the technique, known today as pinch-analysis. At that time he had joined Imperial Chemical Industries (ICI) where he led practical applications and further method development.

Bodo Linnhoff developed the 'Problem Table', an algorithm for calculating the energy targets and worked out the basis for a calculation of the surface area required, known as 'the spaghetti network'. These algorithms enabled practical application of the technique.

In 1982 he joined University of Manchester Institute of Technology (UMIST, present day University of Manchester) to continue the work. In 1983 he set up a consultation firm known as Linnhoff March International later acquired by KBC Energy Services.

Many refinements have been developed since and used in a wide range of industries, including extension to heat and power systems and non-process situations. Both detailed and simplified (spreadsheet) programs are now available to calculate the energy targets.

In recent years, Pinch analysis has been extended beyond energy applications. It now includes:

- Mass Exchange Networks (El-Halwagi and Manousiouthakis, 1987)

- Water pinch (Yaping Wang and Robin Smith, 1994; Nick Hallale, 2002; Prakash and Shenoy, 2005)

- Hydrogen pinch (Nick Hallale et al., 2003; Agrawal and Shenoy, 2006)

Weaknesses

Classical pinch-analysis primarily calculates the energy costs for the heating and cooling utility. At the pinch point, where the hot and cold streams are the most constrained, large heat exchangers are required to transfer heat between the hot and cold streams. Large heat exchangers entail high investment costs. In order to reduce capital cost, in practice a minimum temperature difference (Δ T) at the pinch point is demanded, e.g., 10 °C. It is possible to estimate the heat exchanger area and capital cost, and hence the optimal Δ T minimum value. However, the cost curve is quite flat and the optimum may be affected by "topology traps". The pinch method is not always appropriate for simple networks or where severe operating constraints exist. Kemp (2006) discusses these aspects in detail.

Recent Developments

The problem of integrating heat between hot and cold streams, and finding the optimal network, in particular in terms of costs, may today be solved with numerical algorithms. The network can be formulated as a so-called mixed integer non-linear programming

(MINLP) problem and solved with an appropriate numerical solver. Nevertheless, large-scale MINLP problems can still be hard to solve for today's numerical algorithms. Alternatively, some attempts were made to formulate the MINLP problems to mixed integer linear problems, where then possible networks are screened and optimized. For simple networks of a few streams and heat exchangers, hand design methods with simple targeting software are often adequate, and aid the engineer in understanding the process.

Anchoring on thermodynamic principles, Pinch Technology offers a systematic analysis to optimum energy integration in a process. The observed improvements in the process offered by this technique are not due to the involvement of any advanced unit operations, but due the excellent heat integration scheme it offers. The key advantages of pinch technology over conventional design methods are its ability to set an energy target before the commencement of the design. The energy target offers the minimum theoretical energy required for the process under consideration. Pinch Technology provides the thermodynamic rules to ensure that the energy targets are achieved during the heat exchanger network design.

The main strategy of Pinch Technology is to match cold and hot process streams with the help of a Heat Exchangers Network (HEN) so that requirements for externally supplied utilities are reduced. The best design for an energy-efficient HEN results in a tradeoff between the energy recovered and the capital costs involved for energy recovery.

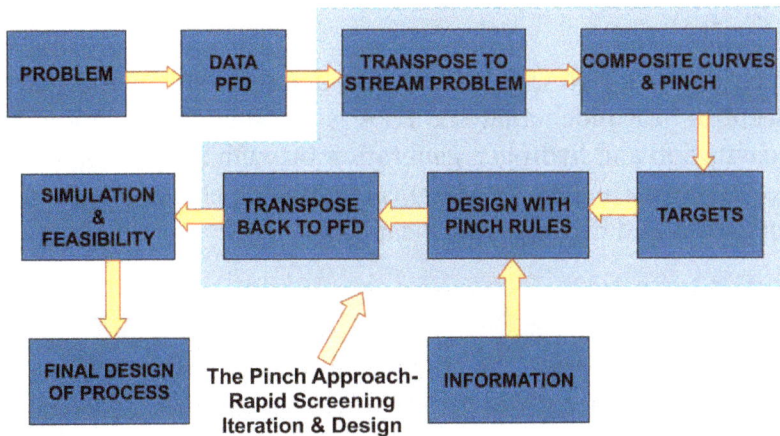

shows the systematic approach required to be taken for process design using Pinch Technology

Key Steps of Pinch Technology

There are four key steps of pinch analysis in the design of heat recovery systems for both new and existing processes:

1) Data Extraction, which involves collecting data for the process and the utility system.

2) Targeting, which establishes targets for design in terms of energy, number of units, area, etc. for best performance

3) Design, where an initial Heat Exchanger Network is established.

4) Optimization, where the initial design is simplified and improved economically.

Water Pinch Analysis

Water pinch analysis (WPA) originates from the concept of heat pinch analysis. WPA is a systematic technique for reducing water consumption and wastewater generation through integration of water-using activities or processes. WPA was first introduced by Wang and Smith. Since then, it has been widely used as a tool for water conservation in industrial process plants. Water Pinch Analysis has recently been applied for urban/domestic buildings. It was extended in 1998 by Nick Hallale at the University of Cape Town, who developed it as a special case of mass exchange networks for capital cost targeting.

Techniques for setting targets for maximum water recovery capable of handling any type of water-using operation including mass-transfer-based and non-mass-transfer based systems include the source and sink composite curves and water cascade analysis (WCA). The source and sink composite curves is a graphical tool for setting water recovery targets as well as for design of water recovery networks.

Hydrogen Pinch

Hydrogen pinch analysis (HPA) is a hydrogen management method that originates from the concept of heat pinch analysis. HPA is a systematic technique for reducing hydrogen consumption and hydrogen generation through integration of hydrogen-using activities or processes in the petrochemical industry, petroleum refineries hydrogen distribution networks and hydrogen purification.

Principle

A mass analysis is done by representing the purity and flowrate for each stream from the hydrogen consumers (sinks), such as hydrotreaters, hydrocrackers, isomerization units and lubricant plants and the hydrogen producers (sources), such as hydrogen plants and naphtha reformers, streams from hydrogen purifiers, membrane reactors, pressure swing adsorption and continuous distillation and off-gas streams from low- or high-pressure separators. The source-demand diagram shows bottlenecks, surplus or shortages. The hydrogen pinch is the purity at which the hydrogen network has neither hydrogen surplus nor deficit.

After the analysis REFOPT from the Centre for Process Integration at The University of Manchester is used as a tool for process integration with which the process is optimized. The methodology was also developed into commercial software by companies such as Linnhoff March and AspenTech. The Aspen product incorporated the work of Nick Hallale and was the first method to consider multiple components, rather than a pseudo-binary mixture of hydrogen and methane.

History

The first assessment based on cost and value composite curves of hydrogen resources of a hydrogen network was proposed by Tower et al. (1996). Alves developed the hydrogen pinch analysis approach based on the concept of heat pinch analysis in 1999. Nick Hallale and Fang Liu extended this original work, adding pressure constraints and mathematical programming for optimisation. This was followed by developments at AspenTech, producing commercial software for industrial application.

Data Validation and Reconciliation

Industrial process data validation and reconciliation, or more briefly, data validation and reconciliation (DVR), is a technology that uses process information and mathematical methods in order to automatically correct measurements in industrial processes. The use of DVR allows for extracting accurate and reliable information about the state of industry processes from raw measurement data and produces a single consistent set of data representing the most likely process operation.

Models, Data and Measurement Errors

Industrial processes, for example chemical or thermodynamic processes in chemical plants, refineries, oil or gas production sites, or power plants, are often represented by two fundamental means:

1. Models that express the general structure of the processes,

2. Data that reflects the state of the processes at a given point in time.

Models can have different levels of detail, for example one can incorporate simple mass or compound conservation balances, or more advanced thermodynamic models including energy conservation laws. Mathematically the model can be expressed by a nonlinear system of equations $F(y) = 0$ in the variables $y = (y_1, ..., y_n)$, which incorporates all the above-mentioned system constraints (for example the mass or heat balances around a unit). A variable could be the temperature or the pressure at a certain place in the plant.

Error types

- Random and systematic errors

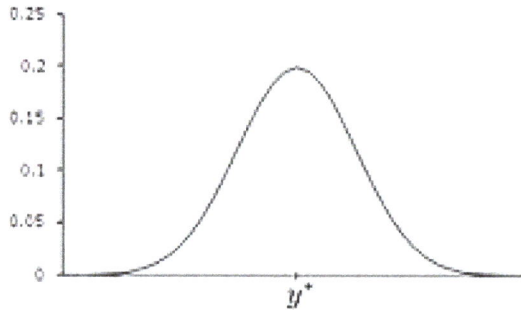

Normally distributed measurements without bias.

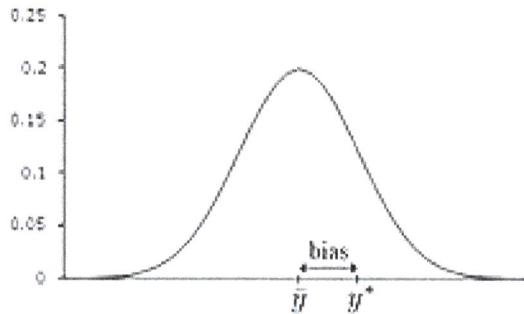

Normally distributed measurements with bias.

Data originates typically from measurements taken at different places throughout the industrial site, for example temperature, pressure, volumetric flow rate measurements etc. To understand the basic principles of DVR, it is important to first recognize that plant measurements are never 100% correct, i.e. raw measurement y is not a solution of the nonlinear system $F(y) = 0$. When using measurements without correction to generate plant balances, it is common to have incoherencies. Measurement errors can be categorized into two basic types:

1. random errors due to intrinsic sensor accuracy and

2. systematic errors (or gross errors) due to sensor calibration or faulty data transmission.

Random errors means that the measurement y is a random variable with mean y^*, where y^* is the true value that is typically not known. A systematic error on the other hand is characterized by a measurement y which is a random variable with mean $\bar{y}$, which is not equal to the true value y^*. For ease in deriving and implementing an optimal estimation solution, and based on arguments that errors are the sum of many factors (so that the Central limit theorem has some effect), data reconciliation assumes these errors are normally distributed.

Other sources of errors when calculating plant balances include process faults such as leaks, unmodeled heat losses, incorrect physical properties or other physical parameters used in equations, and incorrect structure such as unmodeled bypass lines. Other errors include unmodeled plant dynamics such as holdup changes, and other instabilities in plant operations that violate steady state (algebraic) models. Additional dynamic errors arise when measurements and samples are not taken at the same time, especially lab analyses.

The normal practice of using time averages for the data input partly reduces the dynamic problems. However, that does not completely resolve timing inconsistencies for infrequently-sampled data like lab analyses.

This use of average values, like a moving average, acts as a low-pass filter, so high frequency noise is mostly eliminated. The result is that, in practice, data reconciliation is mainly making adjustments to correct systematic errors like biases.

Necessity of Removing Measurement Errors

ISA-95 is the international standard for the integration of enterprise and control systems It asserts that:

Data reconciliation is a serious issue for enterprise-control integration. The data have to be valid to be useful for the enterprise system. The data must often be determined from physical measurements that have associated error factors. This must usually be converted into exact values for the enterprise system. This conversion may require manual, or intelligent reconciliation of the converted values [...]. Systems must be set up to ensure that accurate data are sent to production and from production. Inadvertent operator or clerical errors may result in too much production, too little production, the wrong production, incorrect inventory, or missing inventory.

The amount of information available form plant measurement, data acquisition systems, DCS, and simulation models of a process can be very large, and most of this data may not be of relevance to the Pinch analysis. It is thus necessary to identify and extract only the information that truly captures the relevant sources and sinks (Cold streams) and their interactions with the overall process.

The starting point for a Pinch Technology analysis is to recognize in the process of interest all the process streams that need to be heated and all those that need to be cooled. This necessitates identifying the streams, their flow rates and thermal properties, phase changes, and the temperature ranges through which these must be heated or cooled. This can be performed after mass balances have been completed and temperatures and pressures have been established for the process streams. Energy quantities can be computed by thermodynamic calculations. The above information are generally available in process flow diagrams (PFD). Some heat duties may not be included in the network analysis as these are handled independently without integration. For example, distillation column reboiler heating and condenser cooling may be treated independently of the rest of the heat duties. All the process streams that are to be heated, their temperatures, and enthalpy change rates corresponding to their respective temperature changes or phase changes are then tabulated.

Having obtained a reliable heat and mass balance, the next stage is to extract the hot and cold streams in the form required for pinch analysis. Data extraction can be the most time consuming task of a pinch analysis. It is essential that all the heating, cooling, and phase changes in the process be identified. In existing processes, accurate information may not be readily available, and the engineer may have to go into the field to obtain it. Modeling tools, such as simulation and data reconciliation, can be very helpful in collecting a set of consistent and reliable data.

As mentioned earlier, the key information that needs to be extracted includes the temperature levels of the process streams, and the amount of heat required to bring about desired temperature changes. The enthalpy change rate for each stream is obtained from:

$$\Delta H = mCp\,\Delta T = CP\Delta T$$

Where, ΔH is the enthalpy change rate, m the mass flow rate, Cp the heat capacity, ΔT the temperature change in the stream, and CP the heat capacity flowrate defined as the mCp product. Heat capacity and flowrate are key pieces of information for defining the enthalpy change for a given process stream.

In summary the data required for each process stream include:

- Mass flowrate (kg/s)

- Specific heat capacity (kJ/kg °C)

- Supply and target temperatures (°C), and

- Heat of vaporization for streams with a phase change (kJ/kg).

Additionally, the following information must be collected on utilities and existing heat exchangers

- Existing heat exchanger area (m²)

- Heat transfer coefficient for cold and hot sides of heat exchangers (kW / m² °C).

- Utilities available in the process (water temperature, steam pressure levels, etc

- Marginal utility costs, as opposed to average utility costs.

Data extraction must be preformed carefully as the results of Pinch Analysis strongly depend on this step. A key objective of data extraction is to recognize which parts of the flowsheet are subject to change during the analysis (e.g. possibility of making modifications to the piping, or adding new heat exchangers, possibility of making temperature changes in the process or modifying the utility that heats a given piece of equipment (MP steam instead of HP steam for example), etc). This must be done carefully, as poor data extraction can easily lead to missed opportunities for improved process design. In extreme cases, poor data extraction can falsely present the existing process flow-sheet as optimal in terms of energy efficiency. If, during extraction, all features of the flow-sheet are considered to be fixed, there will clearly be no scope for improvement. In other extreme case, if it does not accept any features of the existing flow-sheet then pinch analysis may over-estimate the potential benefits.

At the beginning of a project it is recommended that all process streams be included in the data extraction. Constraints regarding issues such as distance between operations, operability, control and safety concerns can be incorporated later on. By proceeding in such a fashion, it is possible to have an objective evaluation of the costs of imposing such constraints. A lot of experience is required to ensure that potentially interesting heat-recovery projects are not excluded. Data extraction is a complex issue, and needs expertise.

There are a lot of sector specifics for data extraction. However, heuristic rules have been developed as guidelines. The following are the most relevant:

- Do not mix streams at different temperatures. Direct non-isothermal mixing acts as a heat exchanger. Such mixing may involve cross-pinch heat transfer, and should not become a fixed feature of the design. For example, if the pinch is located at 80°C(hot pinch) and 70°C (cold pinch), mixing a stream at 95°C with a stream at 60°C creates a cross pinch, and will increase the energy targets. The way to extract these streams is to consider them independently, i.e., one stream with a supply temperature of 60°C and the required target temperature, and the other stream with a supply temperature of 95°C. If for process reasons both streams need mixing then these streams should be brought to same temperature say 85 °C before mixing.

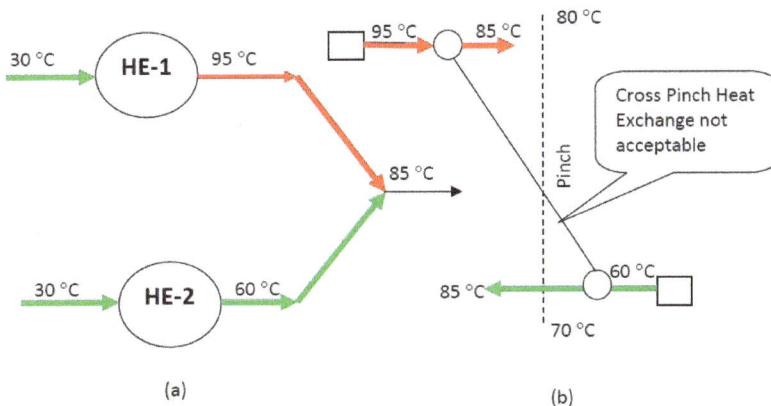

(a)   (b)

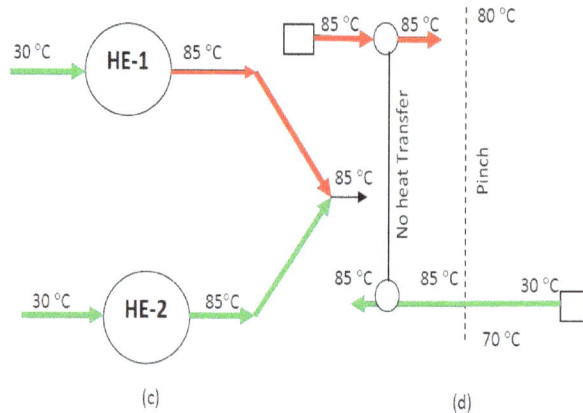

Non isothermal mixing of streams act as a heat exchanger

- For data extraction, the effective stream temperatures are more important than the actual stream temperatures. For example, for a hot stream it is important that at what temperature it is available to exchange heat against cold streams, rather than its actual temperature. Similarly, for cold streams it is the temperature at which heat must be supplied to them is important.

- As CP is a function of temperature, the enthalpy change of some streams is considerably non-linear. The relationship between CP and ΔH with T is given below:

$$Heat\ capacity\ flowrate\ CP = c_0 + c_1T + c_2T^2 + c_3T^3...$$

$$Heat\ load\ \Delta H = c_0T + (c_1T^2/2) + (c_2T^3/3) + (c_3T^4/4)...$$

This is particularly true for phase changing streams such as condensing/vaporising streams. In such a situation, sticking to just one value of CP might lead to inaccurate results. It is more accurate to report the stream in as many "segments" as is required to closely mimic the heating and cooling curve of the stream. However, the position of the extracted stream related to the actual stream curve is also important. In Figure (a), the extracted stream (straight line) is colder than the actual stream near the high temperature end and thus shows false lower temperature. If there is a cold stream with temperature more than the extracted stream but not high enough for the actual stream, an infeasible match will result. Figure (b) features highly safe side approximation because the actual cold stream is colder than each of the three extracted segments. The Figure (c) shows more realistic linearization on the safe side and meets the following rules:

I.   The actual hot stream must be hotter than the extracted hot stream

II.   The actual cold stream must be colder than the extracted cold stream

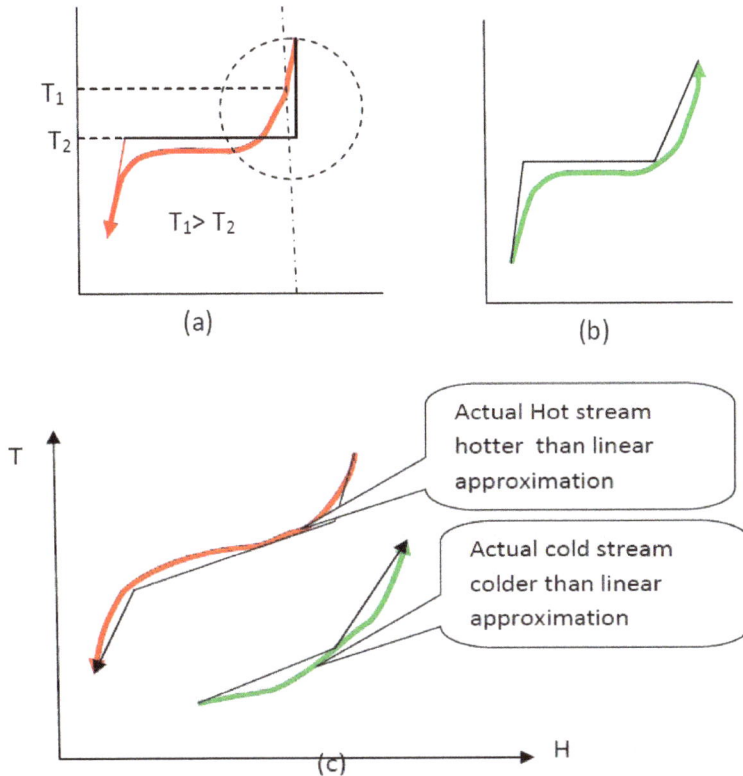

Stream linear approximation, a) and b) may lead to infeasibility where(c) is safe side linear approximation

- Do not include utility streams (steam, cooling water, refrigerant, cooling air, etc.) that can in principle be replaced by any other stream (process or utility) in the process data unless these are involved directly in the process or these cannot be replaced. An example of a utility is cooling water used in an heat exchanger. Since the cooling water can be replaced by air cooling, refrigerant cooling or process stream heating, this should not be extracted. Further, when steam is used in a shift reactor to enhance the shift process, it is not a utility in true sense as it is used as a reactant and cannot be substituted by other stream. Another example that is not so straight forward is when direct steam is used in the reboiler of a distillation column to heat. If it is used just for heating purposes and can be replaced by any other hot utility such as hot oil, steam of different pressure or some other hot utility such as DOWTHERM, it would be treated as a utility and should not be extracted. However, if the reboiling is through direct injection of steam, then the steam is not a utility and should be extracted as part of the process. As one of the aims of using Pinch analysis is to reduce the usage of utilities. Therefore, if utility streams are extracted in a similar way that of process streams, these will be considered as fixed requirements and no opportunities for reduction in utility use will be achieved. In some cases, utility streams can be included because it is not practical to replace them by any form of heat recovery. For example, this is often the case for steam dryers, ejectors and turbine drives.

- Do not consider the existing plant layout. When selecting the inlet and outlet parameters for a process stream, existing heat exchange equipment and plant topology should not be taken into account at first. True utility targets (for cooling and heating) should be set regardless of the existing plant layout. Current plant energy consumption can then be compared with minimum energy targets. In retrofit of existing facilities, once these targets have been determined, plant layout (existing heat exchangers and piping, distances, etc) needs to be taken into account in order to identify practical and cost-effective ways to approach these targets.

- The temperature, pressure and enthalpy conditions of some streams within a process are open to change within certain limits. Such stream data are termed as "soft" data. Soft data should ideally be extracted such that the overall process energy requirement is minimized. For this, the (+)/(-) principle of Pinch Analysis for process modifications is applied. Thus one should identify hard and soft constraints on temperature, pressure and enthalpy levels. For example, a hard constraint would be the inlet temperature of a reactor that cannot be changed in any way, while a soft constraint would be the discharged temperature of a product going to a storage, for which the target temperature is often flexible as the temperature of the stream going into a storage can be within a substantial temperature range.

An example of soft data extraction is shown below:

It is helpful to view the preliminary set of composite curves before to decide upon how best to extract the "soft data" as shown in Figure. Figure (b) depicts a product stream leaving the process boundary at temperature T_1 for product storage. Figure (a) shows the preliminary composite curves for the overall process. T_1 being greater than the pinch temperature (T_{pinch}), useful heat can still be extracted from the stream up to T_{pinch}. This will further reduce the hot utility requirement based on the (+)/(-) principle. The appropriate data extraction, therefore, in this case is pinch temperature.

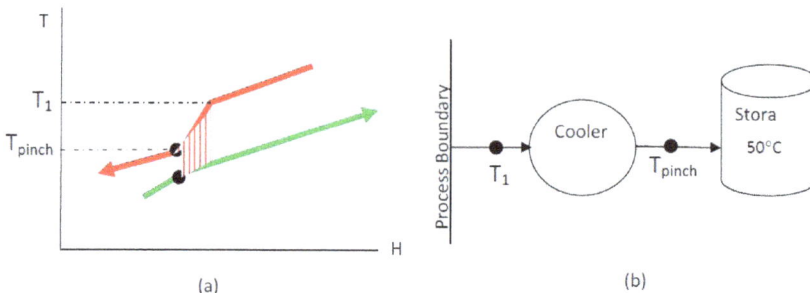

(a) & (b) Extraction of soft stream data

- While extracting data, latent heat loads need some care. Theoretically, these can be considered to be streams with a finite heat load at a fixed temperature, for such streams the heat capacity flowrate CP (CP = $\Delta H/\Delta T$ and $\Delta T=0$) is infinite. In practice, some targeting softwares cannot handle this and the target temperature has to be set slightly different to the supply temperature – say by 0.1°C. In such cases, keep the supply temperature at its original value and alter the target temperature – upwards for a cold (vaporising) stream, downwards for a hot (condensing) stream. The reason is that the pinch is always caused by a stream beginning, so this will ensure that the pinch temperature is still exactly correct.

- Effective temperatures: While extracting stream data from a PFD for hot and cold streams care must be taken to present the available heat at its effective temperature. Let us consider a part of the process to explain it.

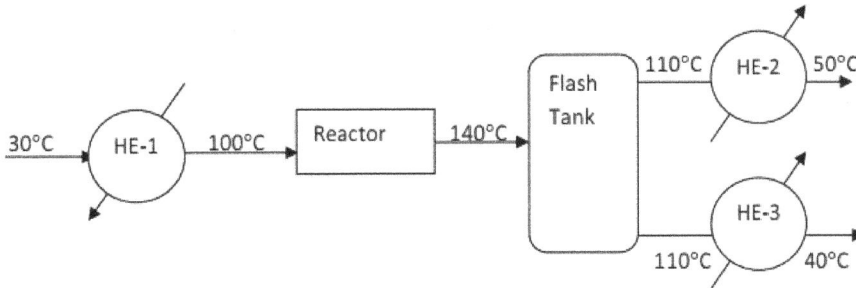

(c) Extraction of effective temperature

The output of the reactor at 140 °C enters the flash tank and cools down to 110 °C. The vapour and liquid leaving the flash tank is 110 °C and are cooled down to 50 °C and 40 °C respectively. Even though the reactor out put is at 140 °C it is not available for exchange of heat. This heat is only available at 110 °C. Thus during data extraction only one cold stream and two hot streams are considered Cold-1 (30 °C to 100 °C), Hot-2 (110 °C to 50° C) and Hot -3 (110 °C to 40 °C).

- Streams that undergo phase-changesand hence have irregular temperature-enthalpy curves can be handled by the addition of extra temperature intervals as indicated by Su.

- Hot streams are defined as those streams which require cooling and in this process transfers heat to cold streams. In general, during this process the temperature of the stream drops. However, the above argument is not valid for the condensing steam which is also a hot stream as it transfers heat to a cold stream.

- Cold streams are those streams which require heating and during this process gains heat. In general, during the above process, its temperature increases. However, the temperature may not increase in some cases when a stream takes heat. For example boiling of a liquid. If the liquid is evaporating due to boiling it is also a cold stream.

Examples of Data Extraction

Example-3.1

The complete flow sheet of a process is given below in Figure and the literal data extracted for stream data table is given in Table.

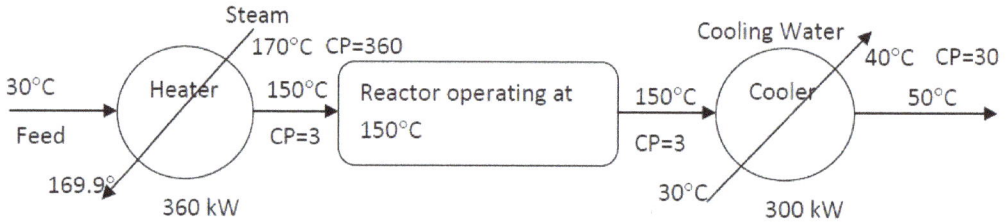

Process flow sheet of a given process

Table: Stream table data for Figure.

Name of the stream	Supply Temperature Ts, °C	Target Temperature Tt, °C	CP kW/°C	ΔH kW
Hot-1	150	50	3	-300
Hot-2	170	169.9	3600	-360
Cold-1	30	150	3	360
Cold-2	30	40	30	300

Minus sign attached with ΔH shows heat given by a stream and plus sing shows heat received by a given stream. It appears from the table that the whole system is in thermal equilibrium and thus energy conservation is not possible. Where as, from the Figure it is clear that internal heat exchange is possible up to the extent of 300 kW and cold utility can be eliminated.

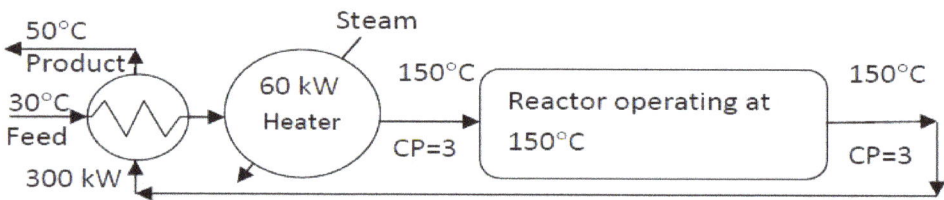

Process flow sheet indicating internal heat

This has happened because we have not followed the rules of data extraction which tells "Do not include utility streams (steam, cooling water, refrigerant, cooling air, etc.) that can in principle be replaced by any other stream (process or utility) in the process data unless these are involved directly in the process or these cannot be replaced." Further it tells that "Therefore, if utility streams are extracted in a similar way that of process streams, these will be considered as fixed requirements and no opportunities for reduction in utility use will be achieved". Hence, utility streams should not be extracted in general.

The proper way of extracting data is shown in Table.

Table: Correct extracted data for Figure.

Name of the stream	Supply Temperature Ts, °C	Target Temperature Tt, °C	CP kW/°C	ΔH, kW
Hot-1	150	50	3	-300
Cold-1	30	150	3	360

Example -3.2

The process flow diagram is shown in Figure and the extracted stream data is shown in Table.

Table: Stream data for problem shown in Figure.

Name of the stream	Supply Temperature Ts, °C	Target Temperature Tt, °C	CP kW/°C
Hot-1	150	30	1.5
Hot-2	170	60	3.0
Cold-1	80	140	4.0
Cold-2	20	135	2.0

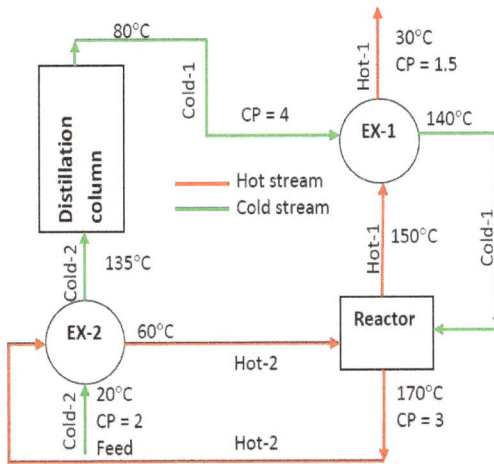

Process flow sheet of a process having distillation column, reactor and heat exchangers

Composite Curves

After the data extraction phase is over, the next step is the creation of hot and cold composite curves. Composite Curves are temperature-enthalpy (T-H) profiles of heat available in the process (through the "hot composite curve") and heat demands in the process (through the "cold composite curve") with the help of graphical representations.

For integration instead of dealing with individual streams an overview of the process is needed. Composite curves are the second step for the integration process. To figure out the heat availability and demand in the process one has to capture the essence of heat load (hot as well as cold streams) integration and heat flow. In other words one should search for a frame work under which integration of heat energy can be performed. To achieve the Pinch Analysis takes help of the first and second law of thermodynamics. The First Law of Thermodynamics provides the energy equation for calculating the enthalpy changes (Δ H) in the streams passing through a heat exchanger and the Second Law determines the direction of heat flow. That is, in natural process, heat energy will only flow in the direction of hot to cold.

Temperature-Enthalpy Diagram

For integration of energy two parameters are most important, amount of energy (given by enthalpy change, Δ H) and direction of energy flow covered by second law of thermodynamics and conserved through maintaining the temperature (T) level sacrosanct. Thus, integration and transfer of energy can be implemented through two parameters namely Δ H and T. The above parameters are shown as two axes of a quadrant below as a 2D representation of a process which can be handled graphically. The plot, as shown in Figure, I s called temperature- enthalpy diagram:

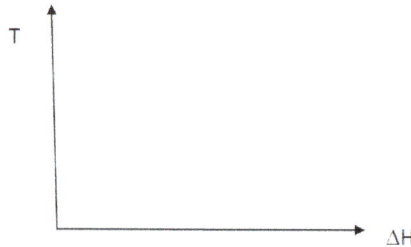

Temperature-enthalpy representation of heat integration

Heat Integration

The heat content Q of a stream (kW) is commonly called its enthalpy H. Further, the differential heat flow dQ, when added to a process stream, will increase its enthalpy (H) by CP dT, where:

CP is heat capacity flowrate (kW/K) (mass flow W (kg/s) x specific heat CP(kJ/kgK)), and dT is the differential temperature change

 Thus, for a stream requiring heating ("cold" stream) from a "supply temperature" (T_s) to a "target temperature" (T_T), the total heat added will be equal to the stream enthalpy change(Δ H) and will be equal to:

$$\Delta H = Q = \int_{T_s}^{T_T} CPdT$$

(1)

Eq. 1 is valid when CP= f(T). In most of the cases CP is a function of T and is given by:

$$Heat\ capacity\ flowrate\ CP = c_0 + c_1 T + c_2 T^2 + c_3 T^3 ... \tag{2}$$

And thus, Δ H becomes,

$$Enthalpy\ change\ \Delta H = c_0 T + (c_1 T^2 / 2) + (c_2 T^3 / 3) + (c_3 T^4 / 4).... \tag{3}$$

The above integration makes Δ H as a nonlinear function of T as shown in Figure. The non-linear temperature –enthalpy behavior can be represented by a series of linear segments.

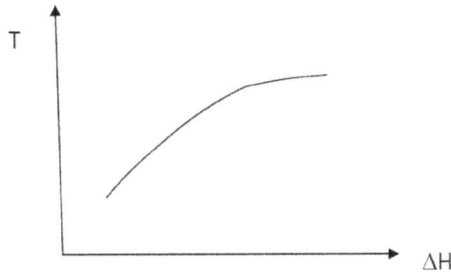

Non-linear representation of a stream.

However, if the value of CP remains constant in the range of investigated temperatures (i.e. T_s to T_T) it can be taken out of the integral sign and thus,

$$H_T - H_S = \Delta H = Q = CP \int_{T_S}^{T_T} dT = CP(T_T - T_S) \tag{4}$$

Where H_S and H_T are enthalpy of the stream taken at T_S and T_T temperatures w.r.t. a common reference enthalpy (H_o) based on the reference temperature, say 0 °C. While dealing with differential enthalpies like (H_T - H_s) the reference enthalpy H_o cancels out.

Or the Eq.4 can be rewritten as:

$$(T_T - T_S) = \Delta T = \frac{1}{CP} \Delta H = \frac{1}{CP}(H_T - H_S) \tag{5}$$

The Eq. 5 resembles with the famous equation of straight line y = mx + c.

where c =0 Where Δ T/ Δ H = 1/CP the slope of the straight line

Thus a hot stream which undergoes a temperature change from T_S to T_T and having a constant heat capacity flowrate as CP can be represented by a straight line in Temperature-enthalpy diagram given in Figure.

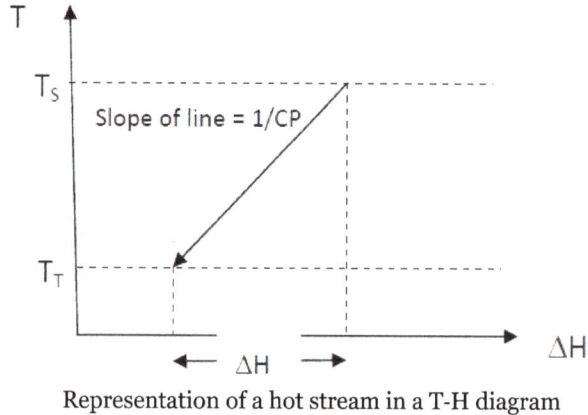

Representation of a hot stream in a T-H diagram

It should be noted that where as, T is treated as an absolute quantity, ΔH is treated as a differential quantity. Due to the differential nature of ΔH it is not dependent on the reference temperature or in other terms any change in reference temperature to evaluate absolute values of H_T and H_S will not affect ΔH.

It will be seen later that the above concept is going to play a vital role in integration of heat loads.

Now let us try to understand the flexibility of the above concept.

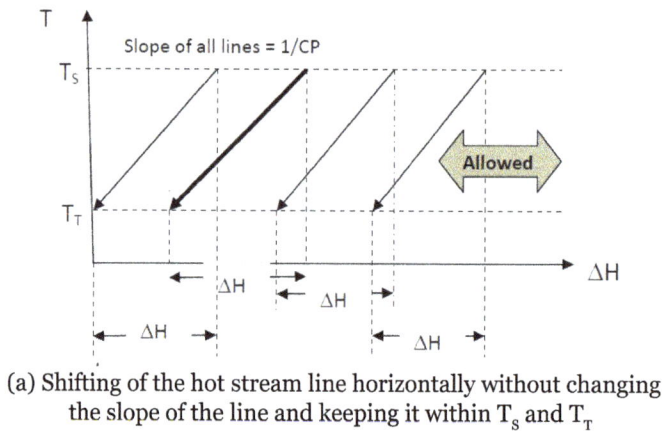

(a) Shifting of the hot stream line horizontally without changing the slope of the line and keeping it within T_S and T_T

From Figure (a) it is clear that the line representing hot stream can be shifted parallel to ΔH axis keeping the slope intact. Further the line should be within the bounds of the temperature T_S and T_T. While doing so it can be observed that the slope is not changing, so the mass flow rate of the stream. This also true for specific heat (Cp) as CP (heat capacity flowrate) is not changing. Further, the temperature interval under which the hot stream is operating is also not changing. Due to the non changing value of CP and temperature interval the value of ΔH is also not changing. In nut shell, it can be concluded that all the vital parameters (CP and temperature levels) are not changing due to the horizontal shifting of the line parallel to ΔH axis. The only thing which is changing is reference enthalpy H_o which in ΔH computation cancels out. Thus it is permitted.

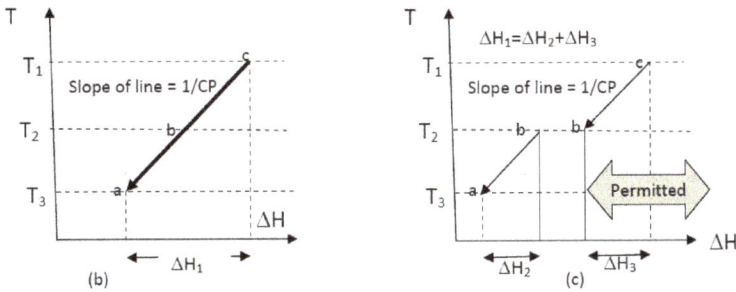

(b) & (c) Splitting of hot stream and shifting it horizontally without changing the slope
of the line and maintaining the temperature intervals

Figure (b) & (c) shows splitting of a hot stream(a-b) based on temperature levels in two
parts such as a-b and b-c. Now both the parts can be translated parallel to ΔH axis.
This is permissible as the slope of the lines are not changing, temperature interval under which these were operating is also not altering and further enthalpy of the stream
is conserved as $\Delta H_1 = \Delta H_2 + \Delta H_3$.

However, it is not true if the line is translated parallel to T axis. Figure shows the shifting of the representative line parallel to T axis for two positions "A" and "B".

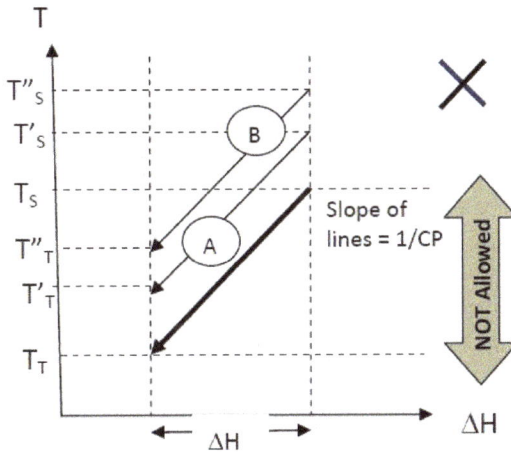

Shifting of the hot stream line vertically without changing the slope
of the line and keeping the value of ΔH constant

From Figure it is clear that when the line is shifted to position "A" the temperature T_T
changes to T'_T and the temperature T_S changes to T'_S. Similarly when the line is shifted
to position "B" the temperatures from original value T_T changes to T''_T and T_S to T''_S.
While doing so, though ΔT remains constant, the temperature levels change. Thus the
values of the important parameter, T, are not conserved. It has been also mentioned
that the values of T are conserved at its original value and any shifting as far as possible
is not accepted at individual stream level as it going to introduce complication in implementation of energy integration. Thus vertical shifting of the line representing hot
stream is not permitted while integration of loads.

Since we are only interested in enthalpy changes of streams, a given stream can be plotted anywhere on the enthalpy axis. Provided it has the same slope and runs between the same supply and target temperatures, then wherever it is drawn on the H-axis, it represents the same stream. Thus, the T/H diagram can effectively be used to represent heat exchange.

Hot Composite Curve

The next step in the integration is to integrate heat loads of two hot streams operating at different temperature levels having some overlap. For this, stream data table as given in Table is considered. This table contains two hot streams.

Table: Two hot stream problem having overlapping temperature range

Name of the stream	Supply Temperature Ts, °C	Target Temperature Tt, °C	CP kW/°C	ΔH kW
Hot-1	125	50	2	150
Hot-2	90	40	6	300

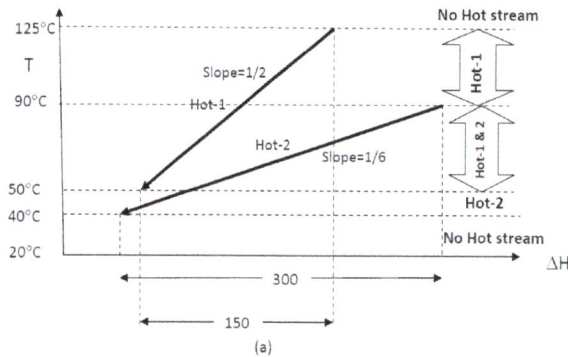

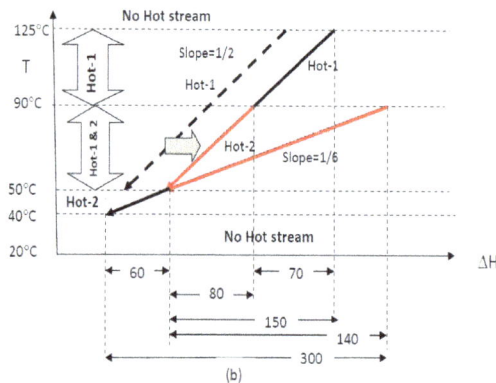

Figure (a) shows the position of Hot-1 and Hot-2 streams in the T-H diagram. From the figure it can be noted that in the temperature levels 40°C-50°C only Hot-2 stream is operating where as for temperature levels 50°C to 90°C both streams(Hot-1 & 2) are operating. Again from 90°C to 125°C only Hot-1 stream is operating. Thus there is a scope of integrating streams Hot-1 and Hot-2 in the temperature levels 50°C to 90°C. Further, it can be noted that ΔH for hot-1 stream from 50°C to 90°C is 80 kW whereas the same for Hot-2 stream is 240 kW. When the load of both the stream are integrated in the above temperature interval these will provide 320kW(240+80). Integration of loads of Hot-1 &2 in the temperature levels 50°C to 90°C is shown in Figure (c).

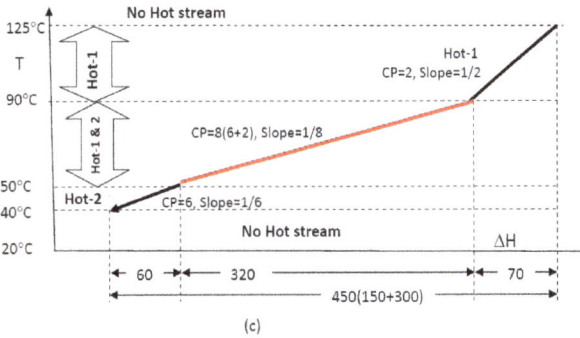

(a),(b) & (c) Heat load integration of two hot stream as given in Table

From the Figure it can be seen that in the temperature levels from 40 °C to 125 °C the total heat available with the hot streams is sum of the heat loads unavailable with Hot-1 and Hot-2 i.e. 450 kW. Further, heat integration of streams within a temperature interval can only be done if these are available in that interval. After integration of loads of hot streams the plot so generated is called Hot composite curve.

To show the integration of hot stream having no overlapping temperature range an example shown in Table is considered.

Table: Two hot stream problem having non-overlapping temperature range

Name of the stream	Supply Temperature Ts, °C	Target Temperature Tt, °C	CP kW/°C	ΔH kW
Hot-1	125	50	2	150
Hot-2	250	130	3	360

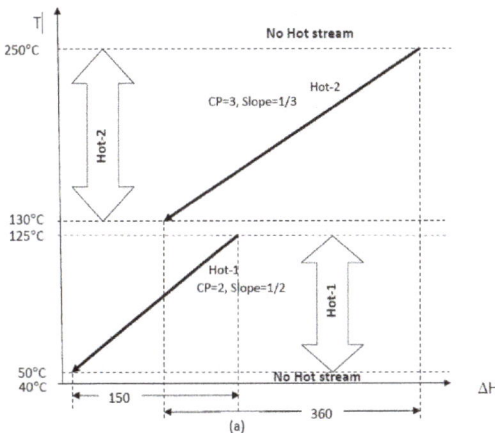

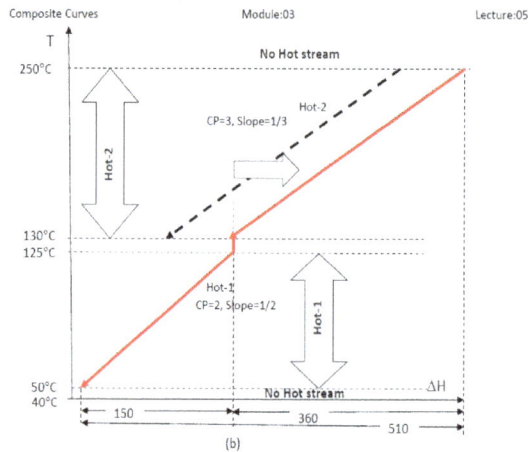

(a) & (b) Heat load integration of two hot stream as given in Table

Figure (b) shows that the Hot-2 stream is moved and brought to the value of ΔH (150 kW) of Hot-1 stream and then the streams(Hot-1 & 2) are joined by a vertical line (parallel to T axis). The enthalpy change for a vertical line is 0 ($\Delta H = 0$) and thus it does not alter the load of the system. However it joins 130 °C level to 125 °C temperature level.

Mathematical foundation for load integration of streams (complete or a part of it) several streams operating under the same temperature interval.

Let us consider there are three hot streams (Hot-1, Hot-2 and Hot-3) operating under same temperature interval, T_1 toT_2 and having heat loads and CP values (within temperature interval of T_1 to T_2) as ΔH_1 & CP_1, ΔH_2 & CP_2 and ΔH_3 & CP_3 respectively. Then,

$$(T_1 - T_2) = \frac{1}{CP_1}\Delta H_1 \qquad for\ Hot-1 \qquad (6)$$

$$(T_1 - T_2) = \frac{1}{CP_2}\Delta H_2 \qquad for\ Hot-2 \qquad (7)$$

$$(T_1 - T_2) = \frac{1}{CP_3}\Delta H_3 \qquad for\ Hot-3 \qquad (8)$$

By manipulating Eqs.6, 7 & 8 and adding one gets Eq. 9,

$$(T_1 - T_2)(CP_1 + CP_2 + CP_3) = \Delta H_1 + \Delta H_2 + \Delta H_3$$

Or

$$\frac{(T_1 - T_2)}{\left(\Delta H_1 + \Delta H_2 + \Delta H_3\right)} = \frac{1}{(CP_1 + CP_2 + CP_3)} \qquad (9)$$

Thus slope of the composite line of all the three Hot streams is $\dfrac{1}{(CP_1 + CP_2 + CP_3)}$

The above method can be generalized for n number of streams.

Instead of dealing with cold streams individually it is desirable to integrate the energy demand of all the cold streams in appropriate temperature intervals and represent these as a composite cold curve. However, in this composite cold curve the temperature levels of each cold stream and heat demand (load) should be preserved. The appropriate temperature ranges are detected from the changes in the T-H plot slopes. If heat capacity flow-rate(CP) are constant, then changes will occur only when streams start or finish. Thus the temperature axis is divided into ranges by the supply and target temperature of streams.

The temperature-enthalpy values (slope of line in T-H diagram) associated with any cold stream can't be changed, however, the relative position of cold streams can be changed by moving them horizontally(parallel to ΔH axis) relative to reach other. This is possible as the reference enthalpy for a cold stream can be changed independently from the reference enthalpy for the other cold streams.

Within each temperature interval, the heat loads of streams (if available) are combined by moving these horizontally to produce a cold composite stream. This cold composite stream, within the said temperature interval, has a CP value that is the sum of the CP values of individual streams present in that temperature interval. Similarly, in a given temperature interval the enthalpy change of the cold composite curve is equal to the sum of the enthalpy changes of all those streams which are present in that temperature interval. The cold composite stream is a virtual single stream that is equivalent to all cold streams present in it, in terms of temperature levels and enthalpy.

Cold Composite Curve

A two stream load integration problem is given in Table. The scope of cold stream load integration is shown in Figure.

Table: Two cold stream problem for load integration

Name of the stream	Supply Temperature Ts, °C	Target Temperature Tt, °C	CP kW/°C	ΔH kW
Cold-1	30	150	2	240
Cold-2	70	120	1.5	75

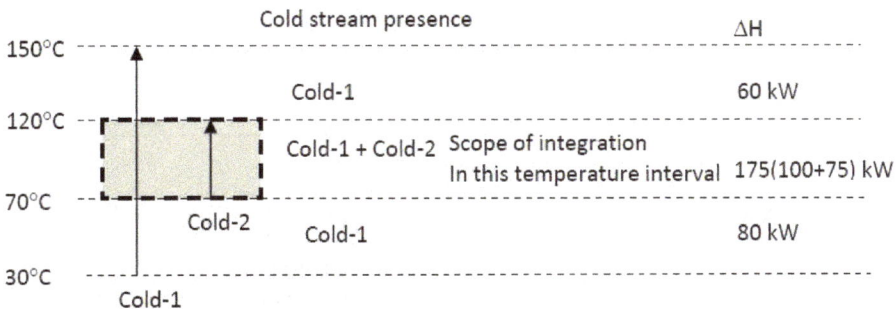

Temperature interval diagram

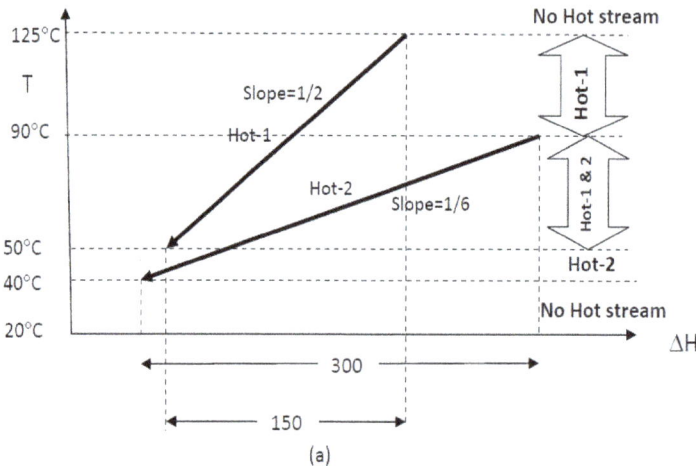

(a)

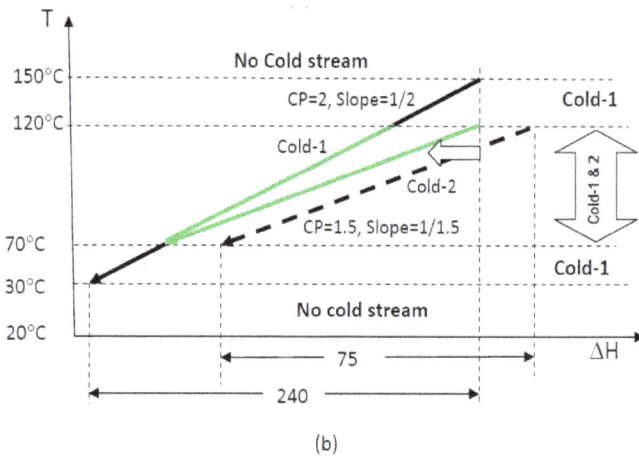

(b)

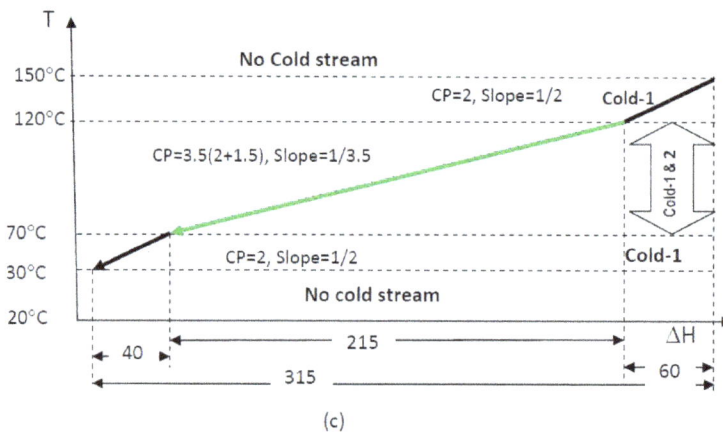

(c)

(a),(b) & (c) Load integration of two cold stream as given in Table

Figure (a),(b) & (c) shows cooling load integration of two cold streams. Figure (c) is the required Cold composite curve. In this curve cold duty integration in the temperature interval 50°C to 120 °C was carried out as in this temperature interval both the cold streams were present.

A four stream load integration problem is given in Table. The scope of cold stream load integration is shown in Figure.

Table: Two cold stream problem for load integration

Name of the stream	Supply Temperature Ts, °C	Target Temperature Tt, °C	CP kW/°C	ΔH kW
Cold-1	30	150	2	240
Cold-2	70	120	4	200
Cold-3	35	60	3	75
Cold-4	130	200	3.5	245
				Σ 760 kW

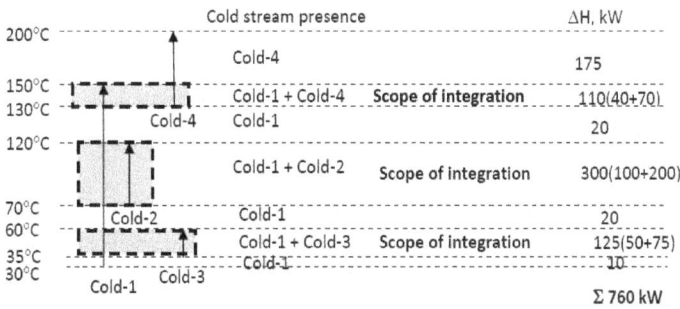

Temperature interval diagram for problem given in table

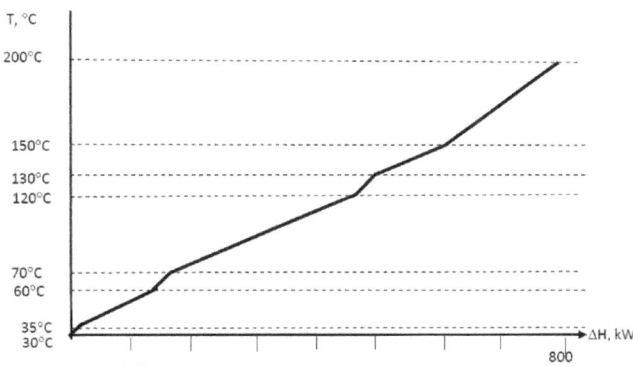

Cold comoside curve for problem of table

Figure shows the cold composite curve for problem given in Table. Further it can be seen that the cold composite curve conserves the heat load of all the individual cold streams (760 kW) and also maintains the supply and target temperatures of all cold streams. Thus it truly represents Cold-1 to Cold-4 streams. Further it can be seen that the slope of the composite curve changes at supply and target temperatures of the individual cold streams.

Pinch analysis is a methodical examination of thermally intensive processes in which all heating and cooling loads (actual or potential) are extracted as temperature/energy flow (T/H) profiles and integrated into composite curves for the whole process and/or site. The following diagrams illustrate the principle of composite curves and show how by combining the hot and cold composites the total potential for heat recovery of the plant can be quantified.

This concept can be explained using a hot and a cold stream as given below:

Table: Two stream problem for prediction of hot and cold utility demand

Name of the stream	Supply Temperature Ts, °C	Target Temperature Tt, °C	CP kW/°C	ΔH kW
Hot-1	210	50	1.5	-240*
Cold-1	90	160	3	210

*-ve sign indicates that the stream is supplying heat

The above system can be pictorially shown below in Figure:

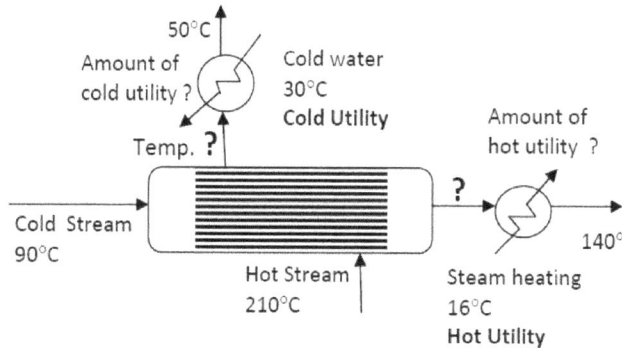

Pictorial view of the problem given in Table

Figure shows the supply and target temperature of hot-1 and cold-1 and position where hot and cold utility are to be supplied. However, it does not provide the amount of cold and hot utilities required and also the amount of internal heat exchange by the Hot-1 and cold-1 streams. Due to incomplete information the exit temperatures of streams from the heat exchanger could not be computed.

From Figure it can be seen that when cold-1 is shifted horizontally towards Hot-1 stream so that it touches the hot-1 stream(position "a") the available ΔT between hot and cold stream at one end becomes zero. This is possible as the reference enthalpy for a cold stream can be changed independently from the reference enthalpy for the hot streams. The point at which the cold stream meets the hot stream is called "Pinch Point". However, in most of the cases "Pinch Point" is not a point but two points called hot pinch point and cold pinch point and are expressed in terms of temperatures. At this point the hot utility (HU) and cold utility(CU) demands are minimum and are equal to 30kW and 60kW respectively. The heat exchange between hot -1 and cold-1 becomes maximum and is equal to 180 kW. Further the total heat exchange including HU and CU becomes 270 kW. However this case is practically infeasible as at $\Delta T = 0$ no heat transfer can take place and area required for heat transfer becomes infinite. So this marks one extreme condition.

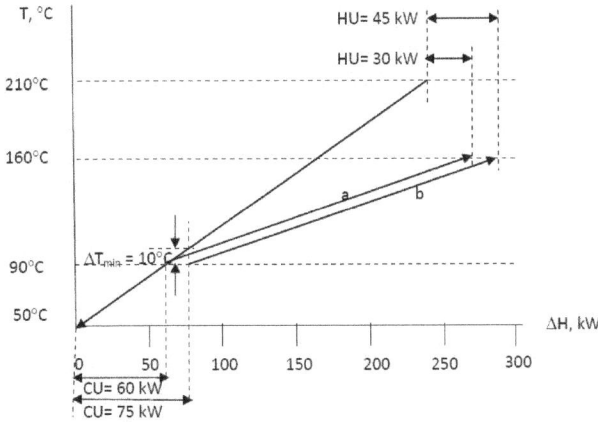

Figure Prediction of hot and cold utility demands as a function of ΔT_{min}

To make the heat transfer feasible the cold-1 stream is shifted parallel to a new position "b" where the vertical distance between its supply temperature end and line representing Hot-1 stream becomes 10°C. This distance is called ΔT_{min}. The value and concept of of ΔT_{min} is important as it fixes the relative position of hot and cold stream (or Hot composite and cold composite). As supply end of cold-1 stream is kept at $\Delta T_{min} = 10^0 C$ from Hot-1 stream the HU and CU demands grow to 45 kW and 75 kW and the internal heat exchange between Hot-1 and Cold-1 becomes 165 kW. The total heat exchange including CU and HU becomes 285 kW. For this case the hot pinch is at 100°C and cold pinch is at 90 °C the difference being equal to ΔT_{min}.

Thus it can be concluded that HU, CU and amount of internal heat exchange is a function of ΔT_{min}. After the above analysis now one is in a position of filling the unknown information of Figure.

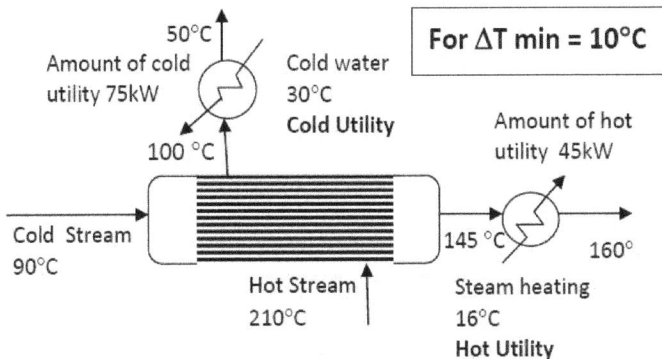

Pictorial view of the solved problem given in Table

After dealing with a two stream problem a comparatively difficult problem as given by stream data of Table is considered for prediction of energy target based on Hot and Cold composite curves and ΔT_{min} equal to 10°C. These curves are shown in Figure.

Table: Four stream problem for load integration and utility prediction for ΔT_{min} equal to 10°C.

Name of the stream	Supply Temperature Ts, °C	Target Temperature Tt, °C	CP kW/°C	ΔH kW
Hot-1	140	50	2	-180
Hot-2	90	40	6	-300
Cold-1	30	150	2	240
Cold-2	70	125	3	165

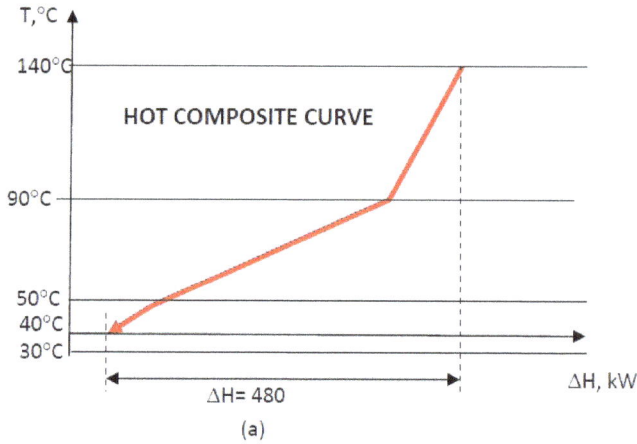

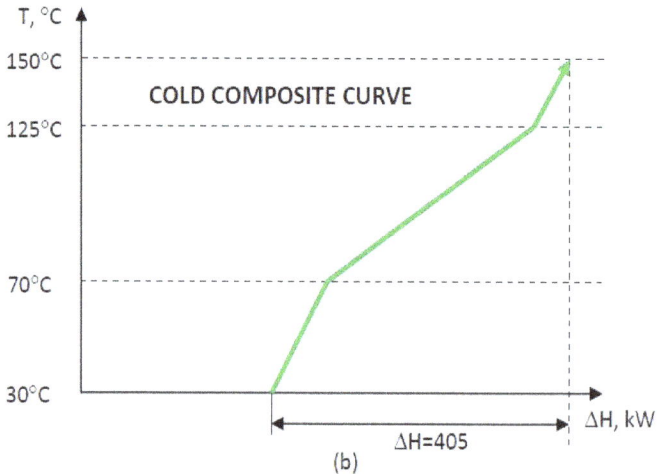

(a) Hot composite curve, (b) Cold composite curve for problem given in Table

Combined Hot and Cold Composite Curves

Once Hot and cold composite curves are created these can be put together to extract useful information regarding energy change and external utility demand.

In Figure the hot and cold composites are plotted on the same T-H axes keeping intact the temperature of both hot and cold composite curves. Hot composite curve contains temperature level of 140°C, 90°C, 50°C and 40 °C(marked by solid lines) and the temperature levels for cold composite curves are 150°C, 125°C, 70°C and 30°C (shown by dotted lines except 30°C as it matches with axis). The cold composite curve is then moved from position "c"(original position) to position "a" where the shortest vertical distance between hot and cold composite curves is 10°C which is the ΔT_{min} in this case. Due to the "kinked" nature of composite curves the point of ΔT_{min} can occur anywhere in the region of internal heat exchange and not just at one end of any composite curve. The temperature-enthalpy values associated with any composite stream can't be changed, however, the relative position of composite streams can be changed by moving them horizontally(parallel to ΔH axis) relative to reach other. This is possible as the reference enthalpy for the cold composite stream can be changed independently from the reference enthalpy for the hot composite streams. For a given value of ΔT_{min} the cold and hot utilities computed are the minimum required values to keep the system in thermal balance.

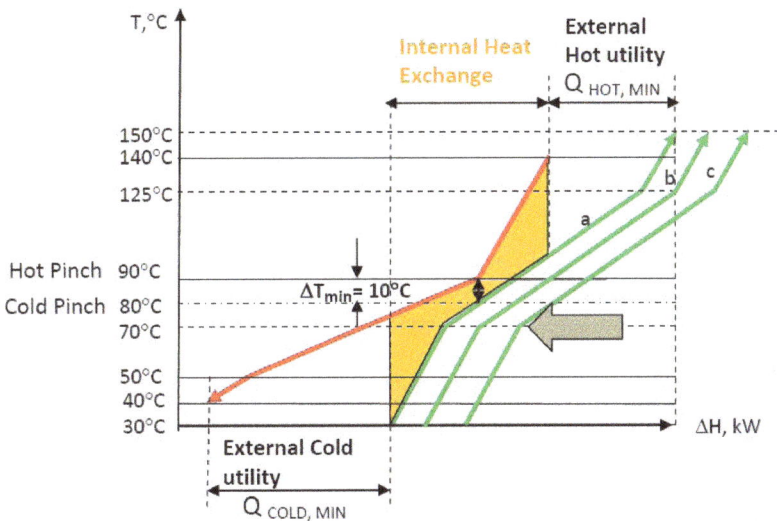

Determination of energy targets utilizing Hot and Cold composite curves

Specifying the hot utility or cold utility or ΔT_{min} fixes the relative position of the two composite curves. The relative position of the two curves is a degree of freedom available to process designer which can be changed by moving these horizontally relative to each other. The shortest vertical distance between two composite curves is called ΔT_{min}. In general, ΔT_{min} occurs at only one point which is called the "Pinch point". This implies that except at or ΔT_{min} in all other places the vertical distance between two composite curves (which is in fact ΔT) will be greater than or ΔT_{min}. Due to the minimum driving force available at ΔT_{min} the design is most constricted here. Further the heat exchanger matches at or ΔT_{min} or near it will have large heat transfer area in comparison to heat exchanger matches which are away from it. When the curves touch each other there is no driving force for heat transfer at one point of the process, which would require infinite heat transfer area and hence infinite capital cost.

It should be noted that the composite curves are made up of many streams. The ΔT_{min} at a certain point between hot and cold composite curves indicate that the all the hot streams present at this point will have a temperature difference of ΔT_{min} (at least at one end of it) with all the cold streams present at that point. So it is possible to design a HEN which will operate the heat exchangers at "Pinch Point" with ΔT values equal to ΔT_{min}. This HEN will demand minimum utility as computed by bringing hot and cold composite curves to a minimum vertical distance of ΔT_{min}.

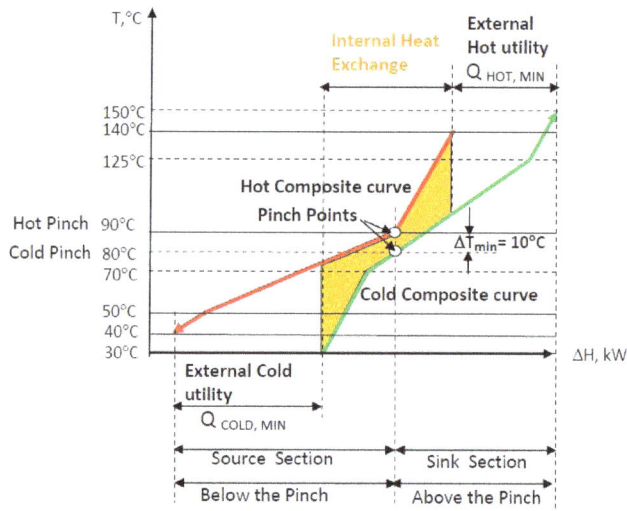

Shows both composite curves put together for problem given in Table

From Figure it can be seen that hot utility demand is 175 kW, cold utility demand is 250 kW and internal heat exchange is 230 kW. This is for ΔT_{min} equal to 10°C. The hot pinch point is at 90°C and cold pinch point is at 80°C.

Main points regarding combined Hot and cold composite curves are:

1. To consider heat recovery from hot stream to cold streams, it is necessary that the complete hot composite curve must be above the cold composite curve for natural heat transfer to take place.

2. The relative position of Hot and cold composite curves can be fixed by fixing Hot utility or Cold utility or ΔT_{min}.

3. The over lapping between hot and cold composite curves determine the maximum extent of internal heat exchange between hot and cold streams.

4. The "over shoot" of cold composite curve (beyond hot composite curve) indicates the minimum amount of external hot utility required for the process.

5. Similarly the "over shoot" of hot composite curve (beyond cold composite curve) determines the minimum external cold utility demand.

6. The Hot and cold utility demands and the amount of internal heat exchange is a function of ΔT_{min}.

7. The ΔT_{min} is defined at pinch point. At this place the vertical distance between two composite curves are minimum and thus it is called ΔT_{min}. The vertical distance between cold and hot composite curves at any place indicates the driving force ΔT which exists at that place. The minimum driving force exists at ΔT_{min}.

8. Pinch divides the complete heat integration problem in two parts Above pinch region (upper pinch) and below pinch region (lower pinch). This is so as in the case of a Maximum Energy recovery design (MER) for a certain value of ΔT_{min} no heat transfer takes place across pinch point. Though physically the heat integration problem appears undivided in terms of the resulting HEN, as far as heat transfer is concerned it is divided in two parts.

9. The Upper pinch area is a heat sink as heat is supplied to this area through external hot utility. The lower pinch area works as a heat source as it transfers heat to a cold utility.

10. The combined hot and cold utility curve helps in plant layout as it indicates which hot stream should be close to which cold stream for minimum piping. It also indicates where the Boiler and Cooling tower should be located which respect to streams.

11. As upper pinch region is a net sink no cold utility should be used here. If a cold utility of amount say X_{cold} is used in this region then the hot utility demand has to be increased to $Q_{H\,min} + X_{cold}$ to satisfy the heat balance of this region. Thus it will increase the hot utility demand. At the same time total consumption of cold utility will also increase to $Q_{C\,min} + X_{cold}$. Thus both cold and hot utilities will increase from their minimum value by an amount X_{cold}. So the penalty will be double in terms of the increased cost of utility as well as heat transfer area required to transfer the heat associated with increased amount of utility.

12. Similarly as lower pinch area is a net heat source and hot utility should not be used here. If a hot utility of amount say X_{hot} is used in this region then the cold utility demand has to be increased to $Q_{C\,min} + X_{hot}$ to satisfy the heat balance of this region. Thus it will increase the total cold utility demand. At the same time total consumption of hot utility will also increase to $Q_{H\,min} + X_{hot}$. Thus both cold

and hot utilities will increase from their minimum value by an amount X_{hot}. So the penalty will be double in terms of the increased cost of utility as well as heat transfer area required to transfer the heat associated with extra amount of utility.

Threshold Problems

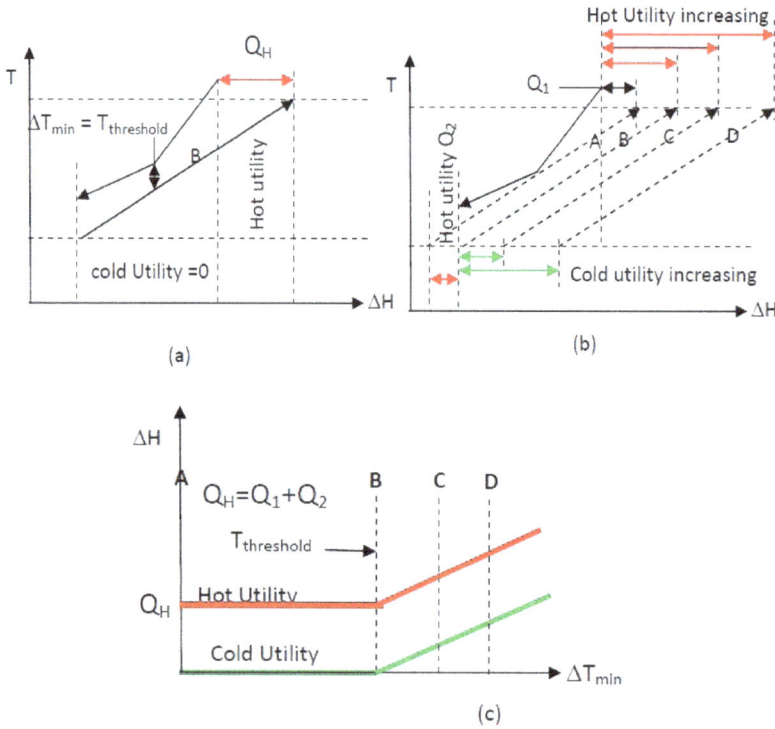

(a),(b) & (c) Threshold problems for different ΔT_{min} values

There are some special category of problems, known as threshold problems, do not have a pinch to divide the problem into two parts. Threshold problems only need a single thermal utility (either hot or cold but not both) over a range of minimum temperature difference ranging from zero to threshold temperature.

For example, in Figure(a) the problem which is at $\Delta T_{min} = T_{threshold}$ requires only one utility i.e. hot utility of Q_H amount. When $\Delta T_{min} \leq T_{threshold}$ as in the case of Figure (b) at positions "A" and "B" the hot utility demand is still Q_H. In the case of "A" two levels of hot utility is required. In the case of position "B" the hot utility demand is Q_1 and Q_2 sum of which is equal to Q_H. Hence it can be concluded that for $\Delta T_{min} \leq T_{threshold}$ the energy demand is not a function of relative positions of Hot and cold composite curves. This is a weakness of pinch analysis. However, this weakness can be supplemented through exergy analysis which provides the lost in-sight. For "a" and "B" positions of cold composite the exergy needs are strongly affected, because of the different temperature levels.

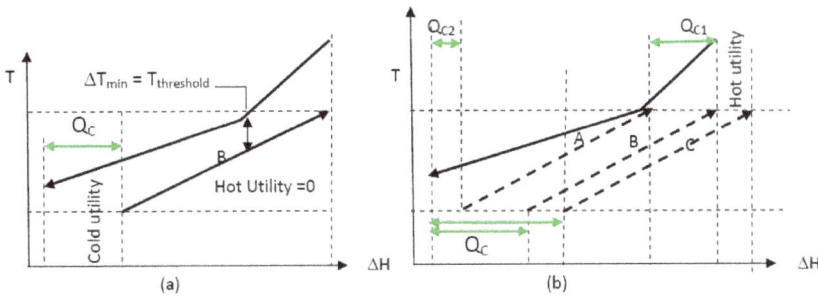

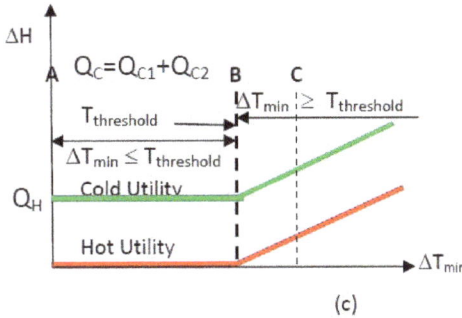

(a),(b) & (c) Threshold problems for different ΔT_{min} values

Figure (a) shows a threshold problem for which hot utility is zero. It only demands cold utility up to $T_{threshold}$. Figure (b) shows the effect of energy demand in terms of cold and hot utilities if the cold composite curve is shifted horizontally to positions "A" and "C". At position "B" which is at $\Delta T_{min} = T_{threshold}$ the hot utility demand is zero whereas the cold utility demand is Q_C. When the cold composite is shifted to position "A" where $\Delta T_{min} < T_{threshold}$ it demands Q_{C1} cold utility at a higher level and Q_{C2} cold utility at a lower level. Where, the sum of Q_{C1} and Q_{C2} being equal to Q_C. For the position "C" where $\Delta T_{min} > T_{threshold}$ the process demands both cold and hot utilities. Thus in this case also for $\Delta T_{min} \leq T_{threshold}$ the cold utility demand is constant and hot utility demand is zero which is shown in Figure (c).

In contrast to the threshold problem Figure shows a "pinched" problem. In this figure both hot and cold utilities are required even if ΔT_{min} is reduced to zero. Further, both the utilities are a function of ΔT_{min}.

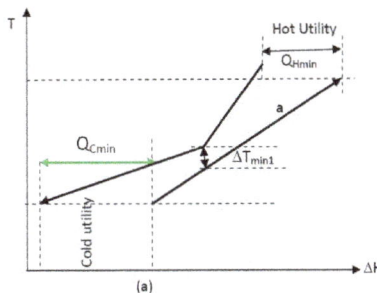

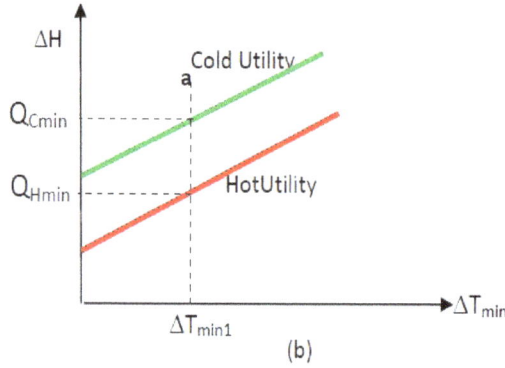

(a) & (b) shows a pinched type problem

Threshold problems can be divided into two broad categories for purpose of design. In the first type, the closest temperature approach between the hot and cold composites is at the "non-utility" end and the curves diverge away from this point. The second type, there is an intermediate near-pinch, which can be identified from the composite curves as a region of close temperature approach.

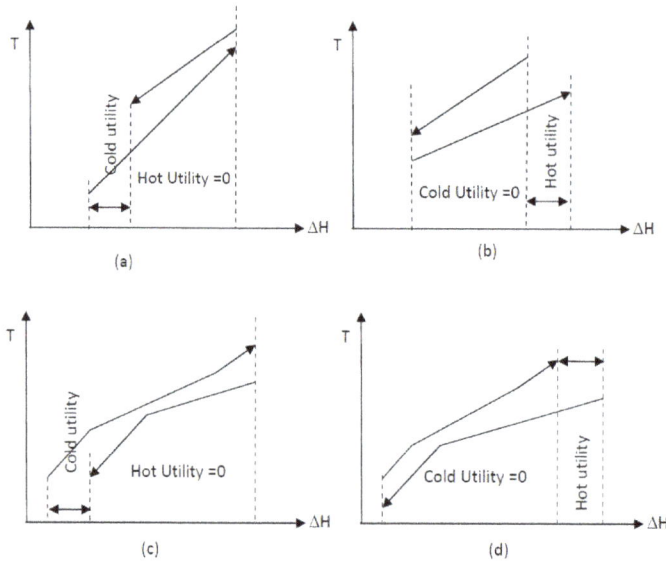

(a),(b),(c) and (d) Different types of threshold problems

Capital-energy Trade-off for Threshold Problems

Figure shows the fixed cost-energy cost trade off as a function of ΔT_{min}. It can be clearly observed that the optimum value either appears when ΔT_{min} is at $T_{threshold}$ or more than $T_{threshold}$. It never happens when $\Delta T_{min} < T_{threshold}$. This is because when $\Delta T_{min} \leq T_{threshold}$ the operating costs are constant since utility demand is constant. Figure (a) shows that optimum is at $T_{threshold}$ where as Figure (b) shows that it is at location where $\Delta T_{min} > T_{threshold}$. In this case there is a demand for both the utilities and thus the problem where there is a pinch. However, in the case demonstrated in Figure (a) there is no pinch.

It can be noted that although threshold problems common and these do not have a process pinch, utility pinches can be introduced in such problems by the induction of multiple utilities.

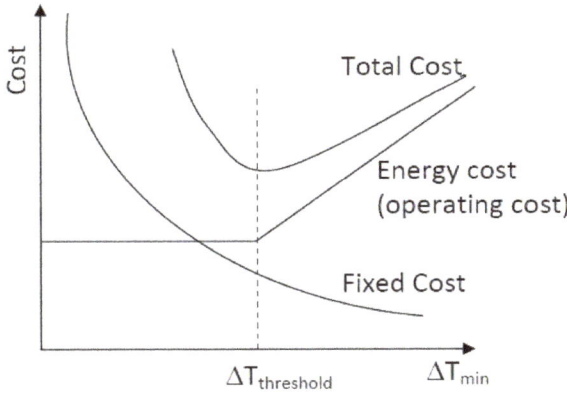

(a) Optimum at ΔT_{min} equal to threshold

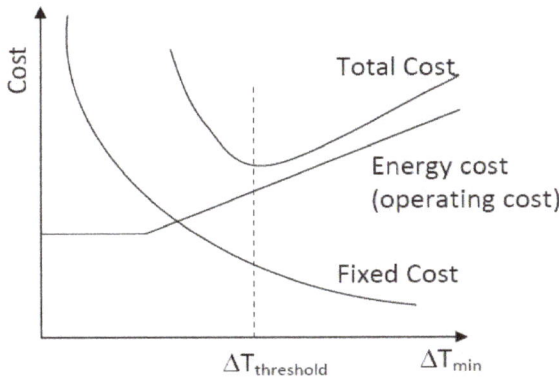

(b) Optimum at ΔT_{min} greater than threshold

Optimum value of capital-energy trade off for threshold problems

Threshold problems are generally handled in design as multiple pinch problems.

Example 01

A four stream threshold problem is given in Table. The Hot and cold composite curves are given in Figure. The hot utility demand for this problem at ΔT_{min} equal to 10 °C is zero and cold utility demand is 239.5 kW. Figure shows the hot and cold utility demand as a function of ΔT_{min} and the value of $T_{threshold}$.

Table: Four stream problem utility prediction for ΔT_{min} equal to 10°C.

Name of the stream	Supply Temperature Ts, °C	Target Temperature Tt, °C	CP kW/°C	ΔH kW
Hot-1	190	55	3.5	-472.5
Hot-2	155	40	1.8	-207
Cold-1	20	140	2	240
Cold-2	70	150	2.5	200

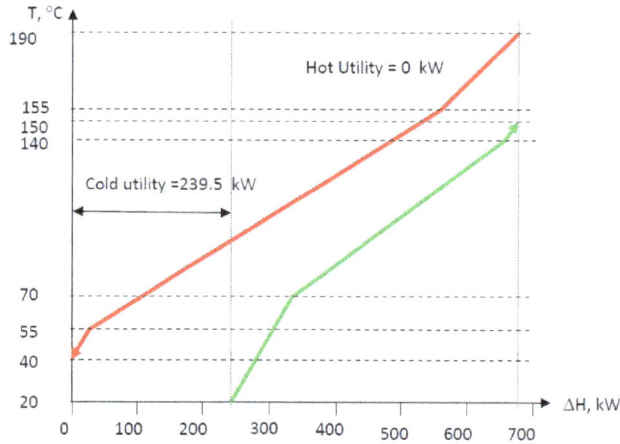

Hot and cold composite curves for the threshold problem given in Table

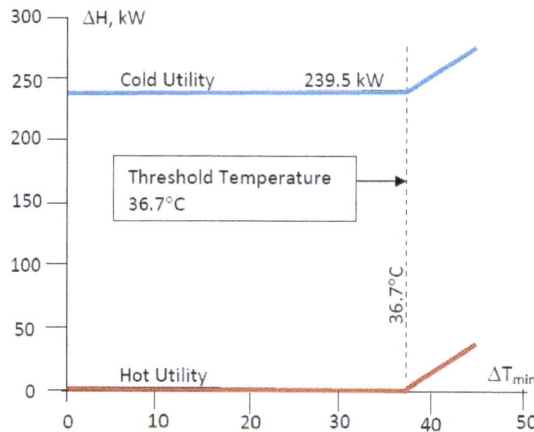

Hot and cold utility demand and threshold temperature for problem given in Table

Example 02

A four stream threshold problem is given in Table. The Hot and cold composite curves are given in Figure. The cold utility demand for this problem at ΔT_{min} equal to 10 °C is zero and hot utility demand is 267 kW. Figure shows the hot and cold utility demand as a function of ΔT_{min} and the value of $T_{threshold}$.

Table: Four stream problem for utility prediction for ΔT_{min} equal to 10°C.

Name of the stream	Supply Temperature Ts, °C	Target Temperature Tt, °C	CP kW/°C	ΔH kW
Hot-1	350	290	3.5	-210
Hot-2	400	290	1.8	-198
Cold-1	150	350	2	400
Cold-2	290	400	2.5	275

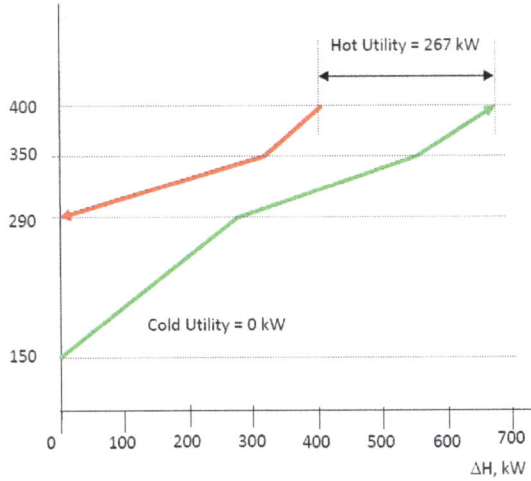

Hot and cold composite curves for the threshold problem given in Table

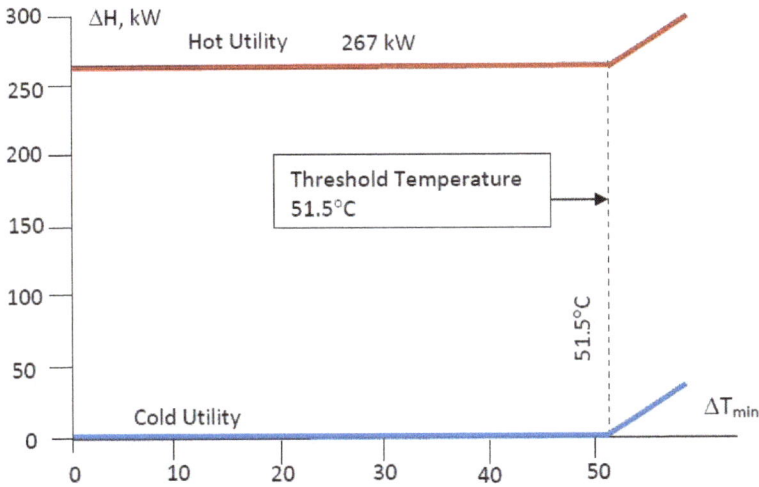

Hot and cold utility demand and threshold temperature for problem given in Table

References

- Ebrahim, M.; Kawari, Al- (2000). "Pinch technology: an efficient tool for chemical-plant energy and capital-cost saving". Applied Energy. 65: 45–40. doi:10.1016/S0306-2619(99)00057-4

- Kemp, I.C. (2006). Pinch Analysis and Process Integration: A User Guide on Process Integration for the Efficient Use of Energy, 2nd edition. Includes spreadsheet software. Butterworth-Heinemann. ISBN 0-7506-8260-4. (1st edition: Linnhoff et al., 1982)

- Manan, Z. A., Foo, C. Y. and Tan, Y. L. (2004). Targeting the Minimum Water Flowrate Using Water Cascade Analysis Technique, AIChE Journal, Volume 50, No. 12, 2004

- Furman, Kevin C.; Sahinidis, Nikolaos V. (2002-03-09). "A Critical Review and Annotated Bibliography for Heat Exchanger Network Synthesis in the 20th Century". Industrial & Engineering Chemistry Research. 41 (10): 2335–2370. doi:10.1021/ie010389e

- Shenoy, U.V. (1995). "Heat Exchanger Network Synthesis: Process Optimization by Energy and Resource Analysis". Includes two computer disks. Gulf Publishing Company, Houston, TX, USA. ISBN 0-88415-391-6

- Wan Alwi, S. R. and Manan, Z. A. (2008). Generic Graphical Technique for Simultaneous Targeting and Design of Water Networks . Ind. Eng. Chem. Res. 47 (8): 2762–2777. doi:10.1021/ie0714870

Pinch Design Method: An Overview

Pinch design methods are very useful as it allows engineers to incorporate real world problems that occur in industries. Pinch analysis is used for the design of heat exchange networks (HENs). The aspects elucidated in this chapter are of vital importance, and provide a better understanding of chemical process engineering.

Rules of Pinch Design Method

The typical heat exchanger network (HEN) synthesis problem can be stated as follows:

For NH hot streams (to be cooled) and NC, cold streams (to be heated) with given supply temperatures, target temperatures, heat capacities, flow rates and film heat transfer coefficients, synthesize the HEN with least total annual cost.

To minimize total annual cost, the synthesis problem should consider all capital cost factors [heat transfer area, number of units (exchangers, heaters and coolers), materials of construction, type of exchanger, etc.] and all operating cost factors (amount of hot and cold utilities, pressure drop and piping cost) simultaneously. Thus the synthesis of a HEN is a large, combinatorial, nonlinear problem.

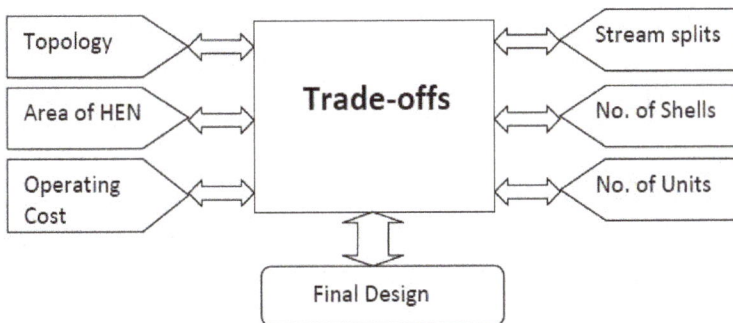

Trade-offs for HEN synthesis

Therefore the decision variables determining the optimal design of a heat exchanger network should be expanded to include the economic trade-offs between units, area, energy, shells, stream splits and topology (operability) as summarized in Figure. The economic trade-off between the first five factors, viz. units, area, shells, energy and stream splits, can be readily quantified. However, it is very difficult to associate a cost with the topology. Ease of operability is a highly desirable goal for any process design but slightly different

Pinch Technology is a widely used technique for the grass-roots design of HENs because of the insight it provides which allows the design engineer to easily incorporate real plant situations for industrial scale problems. Using Pinch Technology one can understand fundamentally how, in heat recovery, the exchanger size and type, F_T factor, number of units, number of shells, area, pressure drop and other aspects fundamentally connect to network structure, operability and energy cost. With Pinch Technology, it is easier to set targets, prior to design, for acceptable payback times in revamp or debottlenecking projects, and to predict initial nearly optimal network and equipment trade-offs from basic principles. Current versions of the Pinch Design Method (PDM) for HEN synthesis yield improved solutions by utilizing the 'Driving Force Plot' (DFP) and 'Remaining Problem Analysis' (RPA).

Though, the pinch design method (PDM) does not guarantee optimal solution. However, it allows a HEN to be synthesized that operates with the minimum energy consumption and at the same time is a good approximation of the optimal network. Furthermore, it gives full control of the design to the engineer, and helps to identify the parameters of the process that are limiting the energy savings.

There are three Important steps in HEN design such as the "targeting" of utility and capital needs prior to synthesis of the HEN, synthesis and optimization of the network and, finally, analysts of network performance under network structure changes and varying operating conditions (i.e. feasibility and resilience analyses)

For these reasons, the pinch design method is a popular and well-established tool for HEN design.

The initial aim is to produce designs using the minimum utility usage with as few capital items, i.e. units, as is compatible. The present section discusses the actual procedure such as the development of "feasibility criteria" which quantify the restrictions placed on design by the pinch, the use of a "tick-off heuristic" to ensure the design is steered towards the fewest possible units and the solution of the "remaining problem" allowing consideration of process constraints and other requirements.

The pinch represents the most constrained region of a design as ΔT_{min} exists between all hot and cold streams at the pinch. As a result the number of feasible matches in this region is severely restricted. Quite often there is a crucial or "essential" match. If this match is not made, this will result in heat transfer across the pinch and thus in increased hot and cold utility usage.

The PDM, therefore

1. recognizes the pinch division

2. starts the design at the pinch developing it separately into two remaining problems.

PDM is completely different from the normal intuitive approach which starts the design at the hot side and moves towards the cold end such intuitive approaches generally violate the pinch. To the contrary, when a design is started at the pinch, initial design decisions are made at the pinch which is the most constrained part of the problem and are less susceptible to difficulties later. At the pinch ΔT_{min} exists between hot and cold streams. This means that the pinch heat exchanger should have a ΔT at pinch end which is equal to ΔT_{min}. If the design is started away from pinch and is moves towards pinch then the follow up matches are likely to violet the pinch or the ΔT_{min} criterion as the pinch point is approached. Thus, if the design is started at pinch and moves away this will not happen. Thus in a PDM, as shown in Figure, hot end and cold end designs start from pinch and moves away.

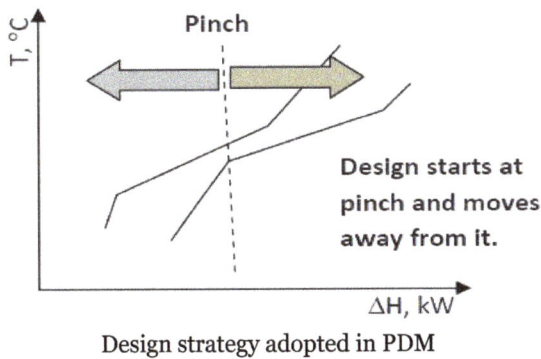

Design strategy adopted in PDM

Feasibility Criteria at the Pinch

There are eight feasibility criteria as given below:

1. Criteria for number of process streams and branches

2. The CP criteria for individual matches

3. The CP difference criteria

4. The CP table

5. The stream tick off rule

6. Remaining Problem Analysis

7. Cyclic matching

8. Rule of CP ratio

The identification of essential matches at the pinch, out of available design options and the stream splitting is so required, is achieved by applying three feasibility criteria to the stream data at the pinch. While developing the above feasibility criteria reference is made to "pinch Exchangers" or "pinch matches" these are defined using Figure.

As shown in Figure pinch exchangers are those exchangers which have the minimum temperature approach, ΔT_{min} on at least one side and at the pinch. It is necessary to identify pinch exchangers in the design as for these exchangers heuristic Pinch Design Rules are implemented sacrosanct.

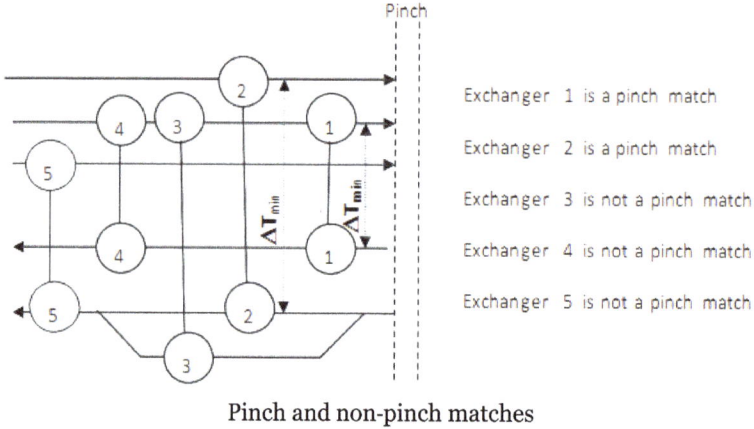

Pinch and non-pinch matches

Criteria for Number of Process Streams and Branches

The first feasibility criterion concerns the stream population at the pinch. The population of hot and cold streams has to be such that it will allow an arrangement of exchangers compatible with minimum utility usage.

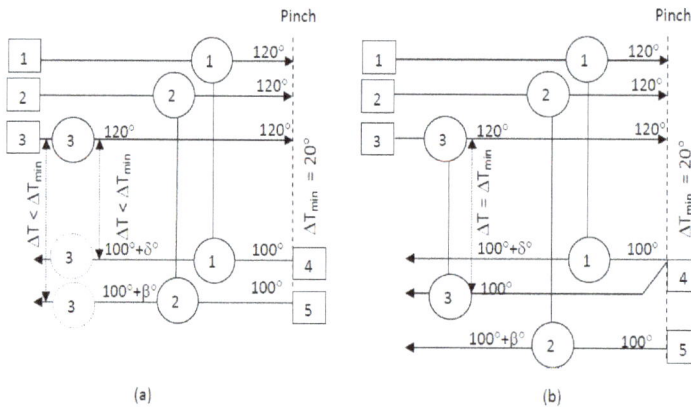

(a) Infeasible design for above pinch region (b) Stream splitting
at above pinch region to get a feasible design

A sample hot end design is shown in Figure. In this problem each process hot stream has to be cooled by process cold streams. As per the pinch rules utility cooling cannot be used in hot end to cool hot streams as it will violate the targeted minimum utility. When attempted to cool the process hot streams by placing pinch matches between hot stream No. 1 and cold stream No. 4 (match exchanger 1) and hot stream No. 2 and cold stream No. 5 (match exchanger 2) as shown in Figure (a), it is possible to cool hot stream 1 and 2. However, after these matches have been made, there is no possibility of cooling hot stream No. 3 using either cold stream 4 or 5 without violating the ΔT_{min} constraint. This is so because the temperature of cold stream after placing the match 1 will rise to a value $100^0 + \delta^0$, where δ^0 will depend on the load of match 1. Thus, if hot stream 3 is matched with cold stream 4 then the ΔT will be less than $\Delta T_{min} (= 20^0 C)$ as can be seen in Figure. The same problem will arise when hot stream 3 is matched with cold stream 5. Thus, it appears that utility cooling would be required to bring the hot stream No.3 to pinch temperature which in fact will increase the cold utility and the design will deviate from minimum utility design.

The above problem can be solved by stream splitting as shown in Figure (b). By split-ting the cold stream No.4 an extra "branch" of the same cold stream is created which allows the hot stream No.3 to match with the "branch" of cold stream 4 without the violation of ΔT_{min} constraint. Thus it can be concluded that a minimum utility design can only be realized if a pinch match can be found for each hot stream. For this to occur inequality given by Eq. 1 must satisfy.

$$NH \leq NC \qquad (1)$$

Where, NH is the number of hot streams or branches and NC is the number of cold streams or branches. Stream splitting may be required to ensure that the inequality in Eq. 1 is satisfied.

On the contrary, diagonally opposite argument applies to below the pinch region. To avoid utility heating at below pinch each cold stream must be brought to the pinch tem-perature by process exchange with hot stream as shown in Figure. As a requirement, a pinch match is required for each cold stream at the pinch and this is only be achieved if Eq. 2 holds good.

$$NH \geq NC \qquad (2)$$

The above problem can also be solved by stream splitting as shown in Figure (b). By splitting the hot stream No.1 an extra "branch" of the same hot stream is created which allows the cold stream No.5 to match with the "branch" of hot stream1 without the vio-lation of ΔT_{min} constraint.

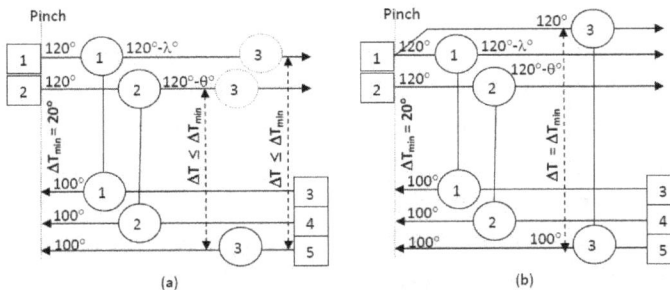

(a) Infeasible design for below pinch region, (b) Stream splitting at below pinch region to get feasible design

The CP Criteria for Individual Matches

The pinch design method starts the design from the pinch where the temperature difference between a pair of hot and cold stream is ΔT_{min}. This constitutes the temperature difference available at one end of the pinch exchanger. Now question is what is the temperature difference at the other end of the pinch exchanger? Whether, it more than ΔT_{min} or less than it. If it is less than ΔT_{min} then it will violet the design as ΔT_{min} is a function of utility requirement and designs are created to achieve a certain utility requirement. Thus to guarantee the utility targets it is necessary that the temperature difference at the other end of pinch exchanger should be more than ΔT_{min}. A search to guarantee this condition gives rise to the CP criteria for individual streams as shown in Figure.

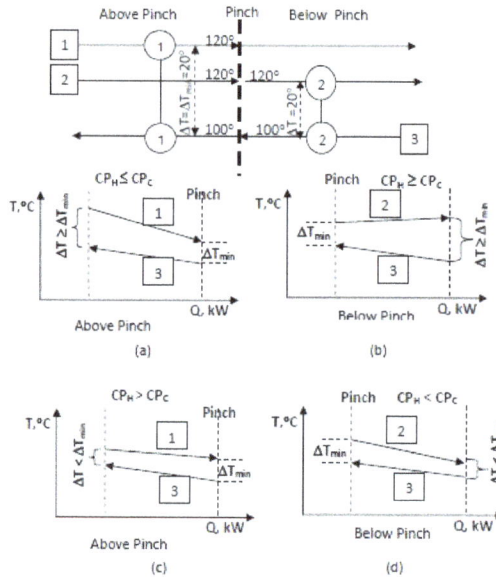

(a) Feasible CP criteria for stream match at above pinch
(b) Feasible CP criteria for stream match at below pinch
(c) Infeasible CP criteria and match above pinch
(d) Infeasible CP criteria and match below pinch

Figure (c) shows the temperature profiles of the match 1 which between hot stream No.1 and cold stream No.3. In this case the CP of stream 1 is greater than CP of cold stream 3. As we know the 1/CP is the slope of the temperature profile (as evident from Composite Curve) the slope of the temperature profile of cold stream 3 will be higher than the temperature profile of hot stream 1. This means that the temperature profiles will converge when these move away from pinch leading to a temperature difference at the other end of the pinch heat exchanger which will be less than ΔT_{min}. This will lead to infeasible matching. To maintain ΔT_{min} at the other end at least the CPs of both the streams should be equal. Under this case both temperature profiles become parallel as 1/CP values of both temperature profiles will be equal. To have a feasible matching, above the pinch (hot end), it is mandatory that CP of hot stream should be lower or equal to CP of cold stream(shown in Figure (a)) necessitating an inequality condition show in Eq 3.

$$CP_H \leq CP_C \quad [above\ Pinch] \qquad (3)$$

Figure (d) shows the infeasible stream matching below the pinch(cold end). In this case also the temperature profiles of hot stream as well as cold stream converge when one moves along the length of heat exchanger away from pinch leading to a temperature difference which is less than ΔT_{min} leading to an infeasible match. This situation will always arise if CP of hot stream will be less than CP of cold stream. This situation can only be avoided if the CP of hot stream will be more than CP of cold stream or at least equal. This will lead to the situation shown in Figure (b) which shows a feasible match. Thus the CP rule for stream matching at below the pinch region is given by Eq. 4.

$$CP_H \geq CP_C \quad [below\ Pinch] \qquad (4)$$

Where CP_H is the heat capacity flowrate of a hot stream or hot stream branch and CP_C is the heat capacity flowrate of a cold stream or cold stream branch.

This should be noted that CP inequality given by Eqs. 3 and 4 are applicable only at pinch or alternately for pinch heat exchangers. These rules can be relaxed considerably for heat exchangers away from pinch. This is because, away from the pinch, temperature driving forces may increase to an extent which may allow matches in which the CP's of the streams matched violate the inequalities. Further, the CP inequalities are reversed when one moves from above the pinch design to below the pinch design.

If it is not possible to create matches fulfilling these inequalities (Eqs. 3 & 4) then it is necessary to change one or more CPs of streams by stream splitting.

The CP Difference

To understand the third feasibility criterion at the pinch it is convenient to define the "CP difference"

For a hot end pinch match

CP difference $= CP_C - CP_H$

For a cold end pinch match

CP difference $= CP_H - CP_C$

Similar equations can be written for differences in the overall sum of hot stream CPs and cold stream CPs

Immediately above the pinch

Overall CP difference $= \sum_1^{NC} CP_C - \sum_1^{NH} CP_H$

Immediately below the pinch

$$\text{Overall CP difference} = \sum_1^{NH} CP_H - \sum_1^{NC} CP_C$$

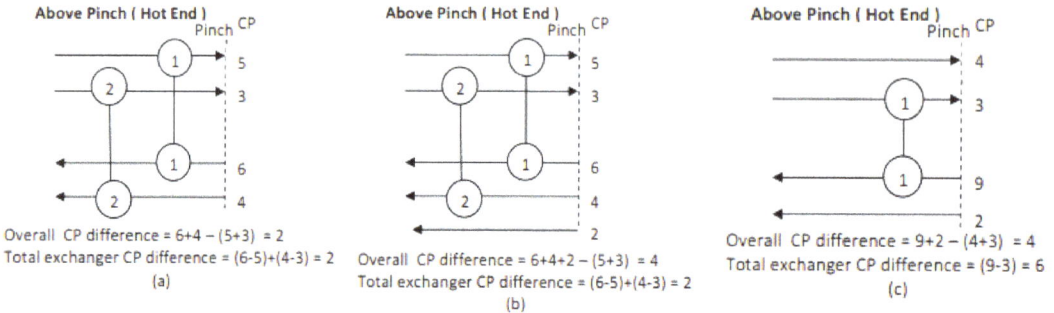

(a) Total pinch exchangers CP difference is equal to overall CP difference, (b) Total pinch exchangers CP difference is less than overall CP difference, (c) Total pinch exchangers CP difference is more than overall CP difference

Figure shows the concept of the CP difference for early identification of matches those are feasible but are not compatible with a feasible overall network.

In Figure (a) shown a hot end pinch design where the sum of the exchanger CP differences equals the overall CP difference meaning that all streams at the pinch are involved through pinch exchangers.

Figure (b) shows a different hot end pinch design where the total pinch exchanger CP differences is less than the overall CP difference meaning that all streams at the pinch are not participating through pinch matches.

Figure (c) shows yet another hot end pinch design where the total pinch exchanger CP differences is more than the overall CP difference. The pinch match (match-1) is feasible as it fulfills inequality Eq. 3 however, it is not compatible as far as feasible network is concerned. In this case the total pinch exchanger CP difference is 6 whereas the overall CP difference is only 4. Thus, the design can't be completed. A match between the remaining hot stream (having CP = 4) and the remaining cold stream (having CP = 2) cannot be matched as per Eq. 3. Thus it can be concluded that for a feasible network design the total pinch exchanger CP differences must always be less or equal to overall CP difference.

Some rules for pinch design methods such as Criteria for number of process streams and branches, CP criteria for individual matches and CP difference criteria. In this section other criteria such as CP table, stream tick off heuristic, stream splitting, remaining Problem Analysis, cyclic matching and CP-rules for minimum area network are discussed.

CP Table

The Pinch Design Method can be benefited with the help of the concept " CP Table". CP tables for the hot and cold end are shown in Figures respectively. The stream problem is shown in Table and the stream population at upper and lower end of pinch is shown in Figure.

CP tables for the problem shown in Table for hot and cold ends are shown in Figs. and table respectively. In these figures hot and cold stream CPs at the pinch are listed separately in decreasing order. The PDM feasibility criteria given by Eqs. 1, 2, 3 and 4 are written at the upper end of the table and the CPs of process streams at the pinch are shown below. The pinch match is shown in the figure by pairing the CPs of a hot and a cold stream. Stream splits are demonstrated by showing branched CPs near the CP of the original stream as can be seen from Figure (c).

Table: A four stream problem for $\Delta T_{min} = 20^0 C$

Stream Number	Stream Type	Heat Capacity Flow Rate	Source Temperature	Target Temperature
		(kW / °C)	(°C)	(°C)
1	HOT	3	160	65
2	HOT	6	90	30
3	COLD	3.5	30	140
4	COLD	4	25	130

(a) CP Table for Table problem (b) &(C) are feasible matches for the problem

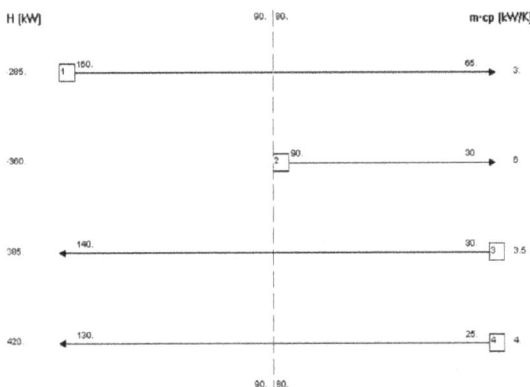

Stream diagram for problem in Table

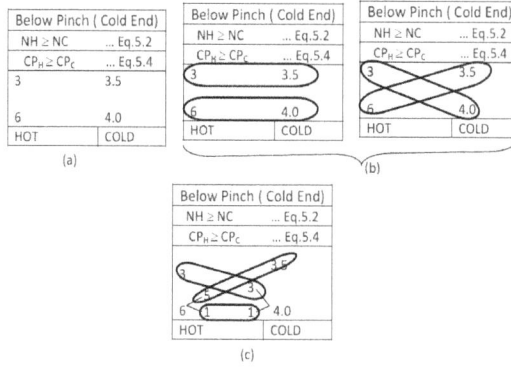

(a) CP Table for Table problem (b) infeasible matching (c) Feasible matching
with one hot and one cold streams split

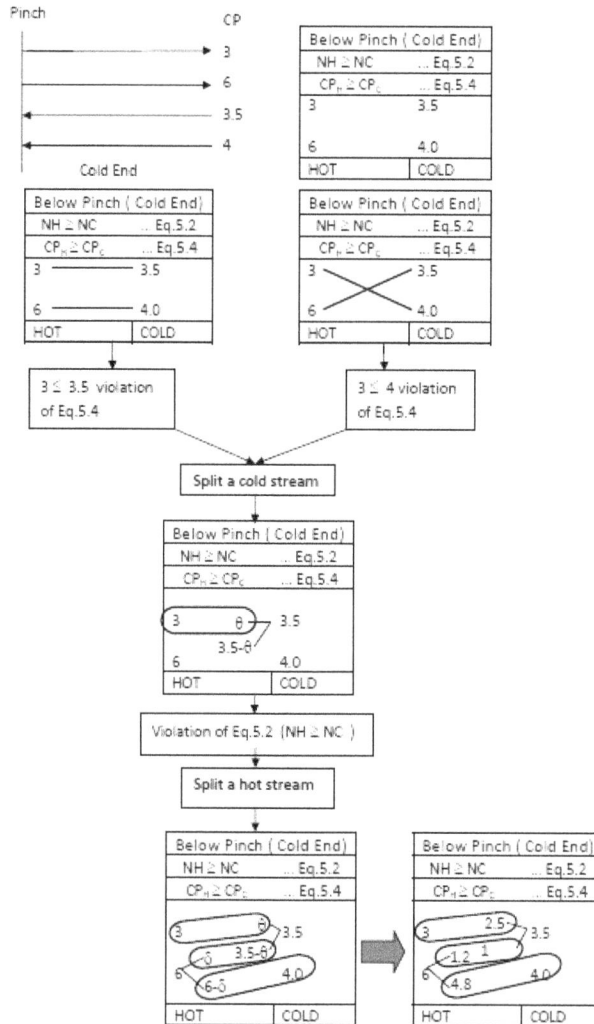

Use of pinch table at cold end to steer stream splitting

Figure (b) shows infeasible pairing of streams having CP values 3 and 3.5 and 3 and 4.

It violates the rule Eq.5.4 though pairing of streams having CPs 6 and 4 as well as 6 and 3.5 is justified as it satisfies the Eq 2.

Figure (c) shows how the hot stream having CP equal to 6 and one cold stream having CP equal to 4 are split and matched in accordance to Eq. 4. This matching also follows the rule given by Eq 2.

To find the values of θ and δ it is recommended that all matches except one be set for CP equality. Thus in Figure if θ is considered as 3, which makes CP_H equal to CP_C for this match, the value of δ comes out to be 0.5. With this all available CP difference ($\sum CP_H - \sum CP_C$ =4.5) is now concentrated in match between hot stream having 9-δ (8.5) and cold stream 4. With this method one quickly identifies the feasible value of δ and θ. However, these values need some tuning based on individual matches considering the "tick off" heuristic or to evenly distribute temperature driving force. If there is no constraints, the split values of CPs are taken so that these are close to the CP values of cold streams these are matching. This minimizes the exchanger area of the HEN. The last CP table of Figure provides a better CP splitting as far as minimization of area is concerned.

For the last CP table one can find that the ratio of $\sum CP_H : \sum CP_C$ is 1.2((6+3)/(4+3.5)). For different matches the CP ratio of hot to cold are 3/2.5, 1.2/1 and 4.8/4. These ratios are equal to 1.2 and are that of the ratio of $\sum CP_H : \sum CP_C$. Besides the values of hot and cold CPs, the temperature range and heat loads of the streams should also be taken into account during splitting. Figure shows placement of heat exchangers after stream splitting shown in Figure.

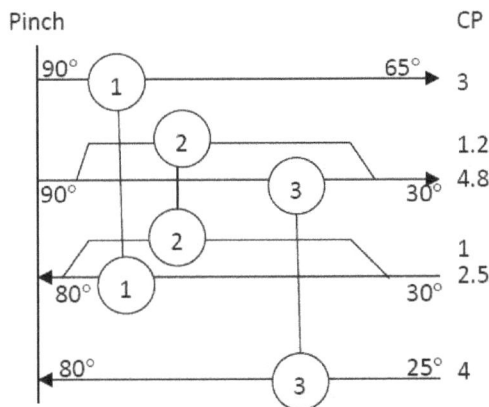

Stream matching after stream splitting of Figure

Figure shows the effect of stream splitting as per $\sum CP_H : \sum CP_C$ as well as loads of heat exchanger for a Hot end design. If the CP of the cold stream is made to split as per heat exchanger loads then the out let temperature of both the braches (before mixing) is same. However this is not true when cold stream CP is made to split according to $\sum CP_H : \sum CP_C$.

	Pinch	CP	ΔH,kW
1 365° ──(1)──	40°	0.4	134
2 260° ──(2)──	40°	0.6	141
3 300° (H) 161° / 130 (2)	30°	2.0	540

T_1 (1) ... T_2

278   132

Cold CP	Split CP	T, °C
Splitting as per $\Sigma CP_H : \Sigma CP_C = (0.4+0.6)/2 = 0.5$		
2.0	0.8	$T_1 = 192.5°$
	1.2	$T_2 = 140°$
Splitting as per Load of HX HX1/ HX2 = 0.98485		
2	0.9924	$T_1 = 161°$
	1.0076	$T2 = 161°$

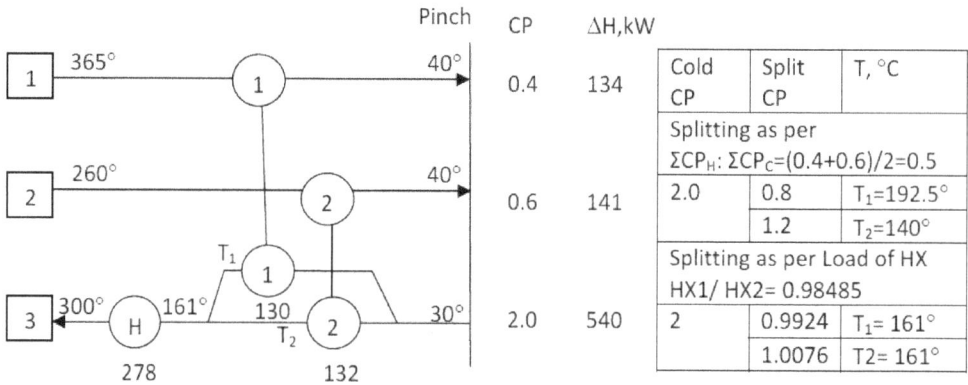

Effect of splitting as per $\sum CP_H : \sum CP_C$ as well as load of heat exchangers

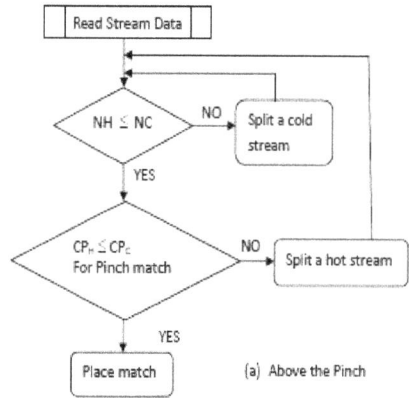

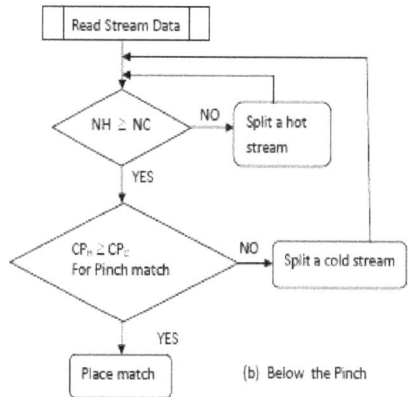

(a) Above the Pinch

(b) Below the Pinch

Stream splitting algorithm for PDM

Figures (a) and (b) show procedure for applying criteria for number of process streams and branches as well as the CP criteria for individual matches. By following the flow chart shown in Figure one can identify essential matches and its options at pinch. The stream splitting need and options can also be identified by the algorithm shown in Figure.

To Summarise this Section on Stream Splitting

- Stream splitting at the pinch is often necessary to achieve an MER design.

- If stream splitting is judged to be undesirable, it can be eliminated by cyclic matching or network relaxation.

- If the designer runs into trouble away from the pinch in applying the ticking-off rule, he can attempt to find a stream split design before resorting to cyclic matching.

- Stream splitting adds complexity to networks as well as flexibility, hence if a nonstream- split, u_{min} solution can not be found, it will normally be preferable to a stream split solution. Note that stream splitting cannot reduce the number of units below the target value.

The "Tick-Off" Heuristic

As soon as the pinch topology has been selected the HEN design should be done in such a manner that its capital cost should be minimum. To achieve the above target the design should be directed towards minimum number of units by using "tick off" heuristic while matches are selected for pinch heat exchanger. The "Tick off" heuristic stems from the minimum number of units target equation.

$$u_{min} = N + L - S \qquad\qquad (5)$$

In general HENs are designed such that S=1 and L=0. So the equation reduces to

$$u_{min} = N - 1 \qquad\qquad (6)$$

The Eq. 5 is satisfied if each match brings one stream to its target temperature or completely utilizes an utility leading to "Tick off" of the exhausted stream or utility and thus the said stream or utility is not a part of the remaining design assignment. Streams at pinch can be "tick off" only if the load of the pinch exchanger dealing these is equal to the smallest load of the streams matched. The required CP inequalities given by Eq. 3 and Eq. 4 will guide in choosing pinch exchanger loads by ticking off stream loads.

The "tick-off" heuristic occasionally punishes the HEN design in terms of increased utility usage. Temperature driving force, required in other parts of HEN, may be consumed considerably by pinch exchangers. For such cases the designer can select option given below:

- decrease the load of the offending pinch match and face the risk of extra units than the u_{min}.

- select a different pinch topology where the tick-off heuristic does not equire vital driving force to be consumed.

From the above discussion it is clear that placing a proper match is difficult and some guiding principles in the form of an analysis is required to steer the design in right direction leading to minimum number of units, area and minimum utility cost. Such an analysis is called remaining problem analysis.

Remaining Problem Analysis

HEN are designed using PDM to achieve the targets estimated by different targeting methods such as utility targeting, no. of shell targeting, area targeting, etc. During design it is required that one should follow the correct path (in terms of correct placement of matches at, near and away from pinch) so that targets are achieved. Remaining problem analysis is one such analysis procedure which uses recursively the Problem table Algorithm or area targeting to steer the design in right direction.

PDM provide a "free hand" to the designer to satisfy process objectives (topology preferences, material safety or other necessary constraints) when designs are carried out away from pinch where temperature driving forces do not restrict the design options. However, the design should not deviate from the targets considerably. Thus the "remaining problem analysis" not only steers the design at pinch but also steers it when the design is away from pinch so that matches are placed in accordance with minimum utility/area objectives.

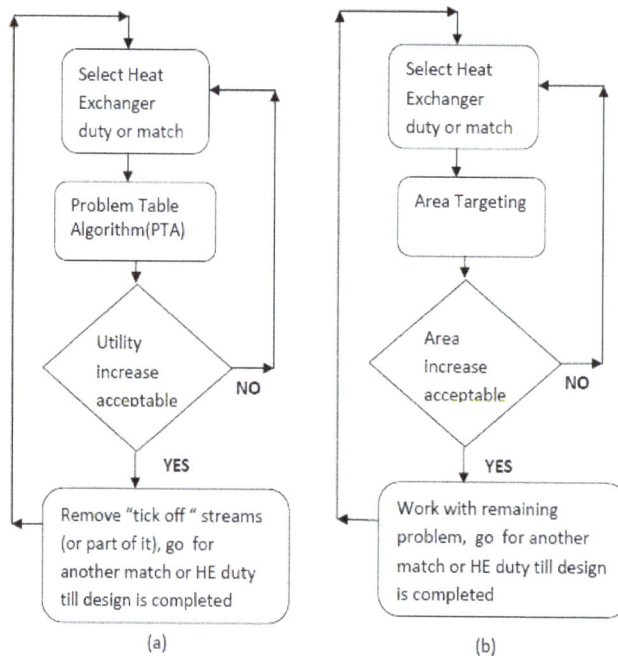

Remaining problem analysis (a) using PTA for utility (b) Using area

As shown in Figure, PTA can be used to check whether a maximized pinch match exchanger load is in accordance with minimum utility usage computed during utility targeting. During a hot end design, PTA applied to above pinch region(hot end) will determine the required hot utility (and no cold utility) as the hot end is a threshold problem which requires no cold utility and that the ΔT_{min}, specified is the threshold value. Now let us analyze a situation where a pinch match has been placed between two streams and its load is estimated using the tick-off heuristic. After this match, the remaining hot and cold streams (or parts of streams) will require a network design to exchange heat. As a part of "remaining problem analysis" now the PTA can be applied to the remaining hot and cold streams(or apart of streams). It may lead to two different estimates of cold and hot utilities. First, the PTA may compute that no utility cooling is required and the utility heating predicted is the same as before for the remaining streams or part of streams (remaining problem). Once the designer gets such a result he knows that he is moving in correct direction as the assigned pinch match load, using the tick-off heuristic, will not penalize the design in terms of increased utility usage. Similar situation can happen for area. The placed pinch match will require some area (say A) and the area targeting with remaining problem will require a certain area (say B). If summation of area "A" and "B" is equal to the minimum area predicted by area targeting or a little more which can be acceptable, then the placement of pinch match is not penalizing the design in terms of area and thus is an acceptable match.

In the second situation PTA may compute that utility cooling will be required for the remaining hot end problem and that the hot utility usage would, increase accordingly. If it is so, then the designer knows that the pinch match load assigned by "tick off" heuristic, is forcing the HEN to use hot utility more than the minimum hot utility. Thus the match is not acceptable and one has to search for another match which will not increase required hot utility for HEN more than the minimum. The same procedure can be used for cold end design which also works as threshold problem requiring no hot utility and that the ΔT_{min}, specified is the threshold value.

Cyclic Matching

Is there any alternative to splitting in the design of HEN? The answer is yes. It is cyclic matching. Stream splitting requires extra pining and extra control equipment to maintain flow in split segments. It may be possible that some streams can not be split as a plant requirement. If one avoid splitting it will be at the cost of penalties on utility as it will violate pinch design rules.

By using the concept of cyclic matching using small heat exchangers a considerable amount of heat can be transferred from hot stream to cold stream without violating the pinch design rules. However, in cyclic matching the size of the heat exchangers will be limited by the violation of ΔT_{min} criterion. Obviously, during cyclic matching the minimum number of units criteria can't be met. Hence, a trade off between the increased cost of the additional exchangers and energy saved has to tried. Theoretically speaking, if an infinite number of infinitely small exchangers are used in cyclically matching then there will not be any energy penalty. Nevertheless, a smaller number of cyclic matches may proved to be sufficient to recover most of the energy as can be seen from the example. Cyclic matching is mostly effective where there is a small pinch region, so that one can move quickly away from pinch region where a considerable number of matches are possible.

The hot end design shown in Figure is considered to show cyclic matching in Figure. Figure (a) shows that if splitting of stream 3 is not considered and matches are placed then hot utility demand rises from 278 to 304.4 and cold utility demand rises from 0 to 26.4. The heat recovery (from hot stream to cold stream) is 89.9%. However, in Figure (b) three matches are made (having a loop) the cold demand utility falls to 9.21 and hot utility demand rises to 287.2. The heat recovery is 96.48%. IN Figure (c) when four matches (two loops)are made then the cold utility demand reduces to 2.58 and hot utility demand falls to 280.58. The heat recovery rises to 99%. Thus with a small number of cyclic matches most of the heat can be recovered. The cyclic matches can be carried out using patching the problem in Excel sheet and taking the load of heat exchangers 2, 3 and 4 as variable in Figure (a), (b) and (c) respectively and adhering to the rule of ΔT_{min} .

What should be the order of matching in a cyclic matching? Temperature is the main criterion for two matches in series. The stream which extends further away from the pinch should be made the "outer" match. Thus, in Figure (a), stream 2 (T_S =260°) should be the match nearer the pinch on stream 3, and stream 1 (T_S =365°) should be the "outer" match way from pinch. However, for matches with a larger number of cycles, the stream with the highest CP should be matched closest to the pinch, as shown in Figure (c).

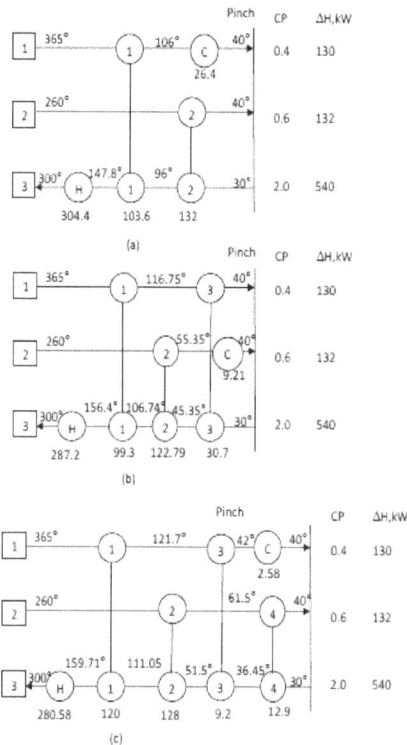

Different cyclic matching for hot end design shown in Figure

The CP-rules for Minimum Area Network

The Pinch Design Method(PDM) requires $CP_H \geq CP_C$ (for cold end design) and $CP_H \geq CP_C$ (for hot end design) for avoiding ΔT_{min} violation, in pinch matches. This rule can be effectively used to obtain near minimum area in the pinch region, where majority of the network area is concentrated. The CP-rules of PDM ensure that the temperature profiles of exchangers at the pinch diverge away from the pinch. The composite curves also diverge away from the pinch. However, the extent of diverge may be different. In fact, if pinch matches have CP-ratios identical to that of the composites, the matches have exactly vertical heat transfer. The design relationship for approaching minimum area around the pinch can therefore be expressed in Equation as:

$$(CP_H / CP_C)_{pinch\,match1} \approx (CP_H / CP_C)_{pinch\,match2} \approx \ldots\ldots\ldots \approx (CP_{Hot\,composite} / CP_{Cold\,composite})_{pinch}$$

Example-1

Table : Four stream problem to exhibit CP rule for minimum area of HEN

Name of the stream	Supply Temperature Ts, °C	Target Temperature Tt, °C	CP MW/°C	ΔH MW
Hot-1	155	55	0.22	22
Hot-2	175	45	0.12	15.6
Cold-1	55	125	0.33	23.1
Cold-2	80	120	0.52	20.8

U= 110 W/m² °C for all matches Cooling Duty = 4.85 MW and Heating duty = 11.15 MW

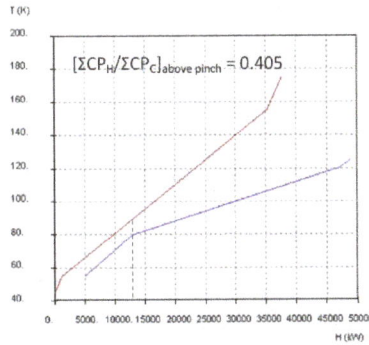

GCC of the example problem

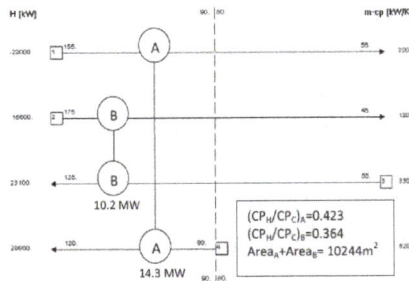

Exchanger placement(topology-1) for above pinch

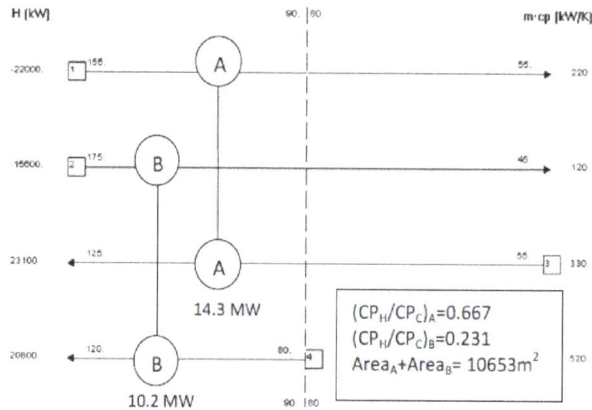

Exchanger placement (topology-2) for above

Example 1 in Figures show two different topologies for a stream set above the pinch. Both networks obey the CP-inequalities as far as basic feasibility is concerned. However, network in Figure has pinch matches with CP-ratios closer to that of the composites and obtains a lower area than network in Figure.

Design Method Summary

The pinch design method (PDM) can be summarized as follows:

- The HEN problem is divided at the pinch into two separate problems.

- The design of the separate problems is started at the pinch moving away from the pinch.

- To satisfy ΔT_{min} criterion, constraints imposed by feasibility criteria on the CP values are satisfied for pinch matches. The stream splitting requirements at the pinch are identified by applying the feasibility criteria.

- The heat loads of exchangers at the pinch are determined using the "tick-off" heuristic to minimize number of units. This heuristic in some cases penalize the design in terms of need for increased utility usage. In such cases a different exchanger topology at the pinch should be preferred or the load on the offending match be reduced.

- Away from the pinch there is exists more freedom regarding the choice of matches due to the availability of sufficient driving force. Thus the designer can put matches based on operability, plant layout or process requirement.

- At pinch for pinch matches the CP rule

$$(CP_H / CP_C)_{pinch\,match1} \approx (CP_H / CP_C)_{pinch\,match1} \approx \ldots \approx (CP_{Hot\,composite} / CP_{Cold\,composite})_{pinch}$$

ensures minimum area of the network

Table: A four stream problem for $\Delta T_{min} = 10^0 C$

Stream Number	Stream Type	Heat Capacity Flow Rate	Source Temperature	Target Temperature
		(kW / °C)	(°C)	(°C)
1	HOT	3	200	65
2	HOT	6	90	30
3	COLD	3.5	30	142
4	COLD	4	25	130

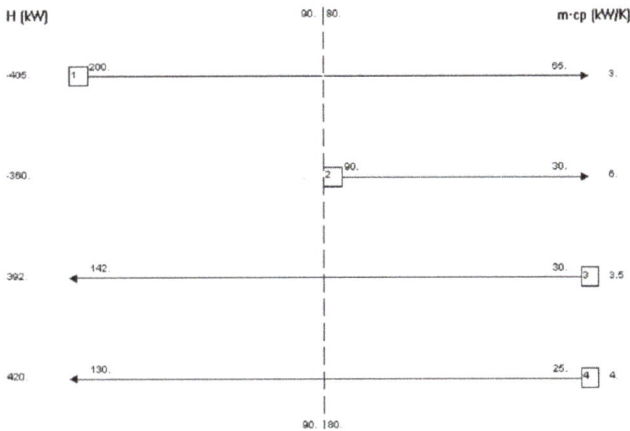

Stream diagram for problem in table

For the given problem: $Q_{H\,min} = 87\,kW$, $Q_{C\,min} = 40\,kW$, Hot Pinch temperature = 90°C and cold pinch temperature = 80°C

Hot End (Above Pinch) Design

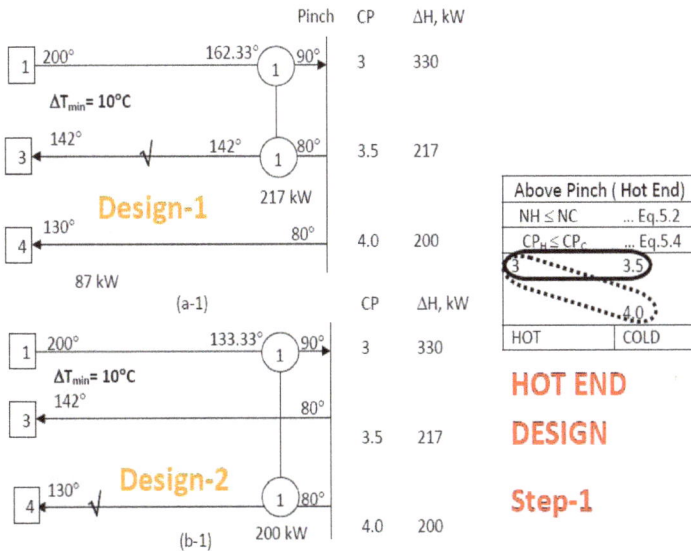

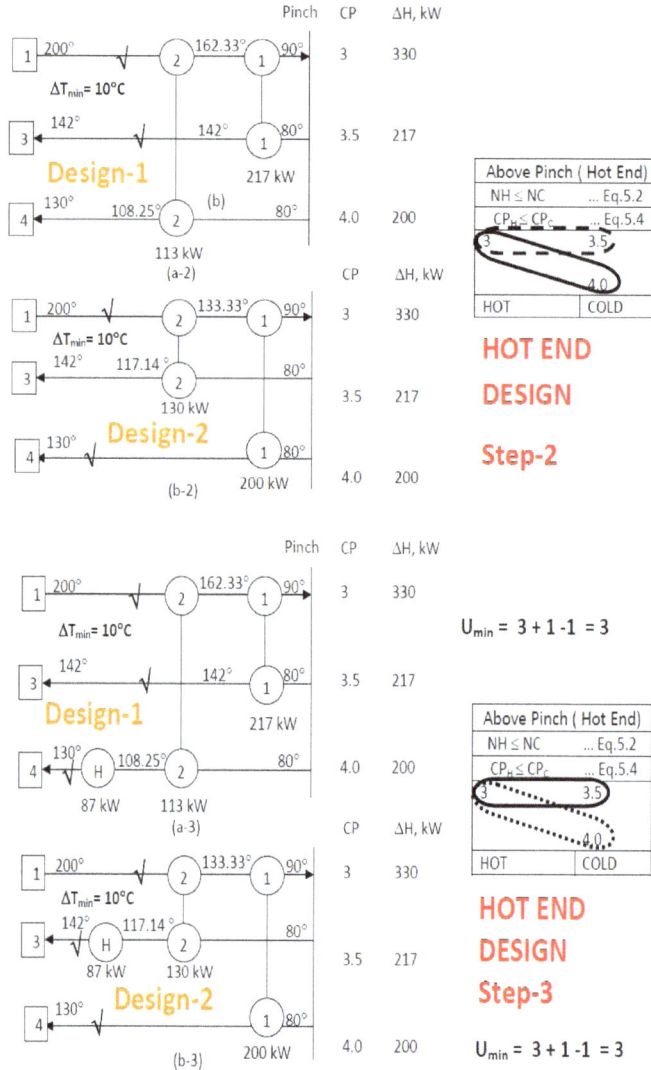

Figure Different steps (a-1) to (a-3) for feasible design-1 and & (b-1) to (b-3)
for feasible design-2 of the Hot end design for problem in Table.

Design-1

Figure (a-1) shows the placement of heat exchanger between stream-1 and stream-3. As the CP of stream-3 is 3.5 and that of stream-1 is 3.0, the CP of stream-3 > to CP of stream-1(required by PDM) and thus, a match can be placed between these streams. Please note that this exchanger is a pinch exchanger. As per the "tick off" rule this match can be maximized to 217 kW so that the stream-3 is ticked off as indicated by the sign "√" placed on the stream. In the CP table this match is shown by a solid line.

The second match which is shown in Figure (a-2) and is not a pinch match, is placed between stream-1 and stream-4 as CP of stream-4 > CP of stream-1. This match is for the remaining duty of stream-1 which is 113 kW. This match is shown in CP table of Figure (a-2) by solid line. With this match stream-1 is "ticked off". Now the only stream whose heat load is not satisfied is stream-4. To satisfy the heat load a heater of 87 kW is placed on this stream. By doing so the stream-4 is ticked off. Further, for this part of problem number of cold stream is greater than number of hot stream as required by PDM and thus no stream splitting is required to complete the design. Figure (a-3) shows the complete hot end design of the problem in Table.

Number of units for this design is equal to 3. This is as per the number of unit target applied to hot end of the problem as shown in Figure (a-3).

Design-2

The design-2 provides an alternate design to design-1 following the PDM showing that many alternate feasible designs can be developed for the same problem. The placement of heat exchanger between stream-1 and stream-3 is shown in Figure (b-1). As the CP of stream-4 is 4 and that of stream-1 is 3.0, the CP of stream-4 > to CP of stream-1 (as required by PDM) and thus, a match is placed. Please note that this exchanger is a pinch exchanger. As per the "tick off" rule this match can be maximized to 200 kW to "tick off" stream-4 as indicated by the sign "√" placed on the stream. In the CP table this match is shown by a dotted line.

The second match, a non pinch exchanger, as shown in Figure (b-2) is placed between stream- 1 and stream-3 as CP of stream-3 > CP of stream-1. This match is for the remaining duty of stream-1 which is 130 kW. This match is shown in CP table of Figure (b-2) by dotted line. With this match stream-1 is "ticked off". Now the only stream whose heat load is not satisfied is stream-3. To satisfy the heat load of stream-3 a heater of 87 kW is placed on this stream. By doing so the stream-3 is also ticked off. Further, for this part of problem number of cold stream is greater than number of hot stream as required by PDM and thus no stream splitting is required to complete the design. Figure (b-3) shows the complete hot end design of the problem in Table. Number of units for this design is equal to 3. This is as per the number of unit target applied to hot end of the problem as shown in Figure (b-3).

The final hot end design is shown in Figure (a-3) for feasible design-1 and in (b-3) for feasible design-2. Bothe designs, design-1 and design-2 are correct as far as PDM is concerned. Both the designs are a part of MER design as these consume the targeted minimum hot utility demands. However these may differ in total area of the heat exchanger network and thus fixed cost.

Cold End (below Pinch) Design

As shown in hot end design, two designs namely design-1 and design-2 have been proposed for cold end design also as shown below. Both designs are feasible as far as PDM is concerned. These also consume the minimum cold utility demand and thus are a part of MER design. However, these designs will require different network area and thus fixed cost.

The design of cold is more complex than hot end as appears from the design. As per the

PDM as no. of hot stream is equal to no. of cold streams, stream splitting is not required. However, the design compulsions required stream splitting of a hot as well as a cold stream. In this case also the design is conducted in different steps starting from step-1 to step- 3.

As per the CP table and PDM only one straight forward match is possible between stream-2 and stream-4. The number of units target shows that $u_{\min} = 4$. Thus the design strategy has to be changed to put 4 matches to maintain heat balance between hot and cold streams as well as cold utility.

Design-1

For design-1 the stream No.3 is split in to two parts to offer a possibility for matching of stream- 1 with a part of stream-3. Thus, stream-3 is split in such a way that one part of it will tick off stream-1. Therefore, stream-3 is split into two parts having CPs equal to 1.5 and 2. The stream having CP equal to 1.5 will offer a load of 75 kW. Now stream-1 is matched with a part of stream-3 having CP equal to 1.5. This ticked off stream-1 as well as a part of stream-3 having CP equal to 1.5 as can be seen from Figure (a-1). This matching is shown in CP table by blue solid line.

As we have only one hot stream (stream-2) is left, the cooler has to be placed on it. Thus a cooler is placed away from the pinch point on this stream as shown in Figure (a-2). The load of the cooler is 40 kW as reported in the problem $\left(Q_{C\min} = 40\,kW\right)$. A simple computation shows that the input temperature to cooler will be 36.66°C.

Now stream-2 is split in two parts to provide balanced heat(after cooler) to satisfy the heat load of stream 4 and split part of stream-3 having CP equal to 2. As the CP of stream 4 is 4.0, one of the split part of the stream-2 should have a CP equal to 4.0 to match CP criterion. Thus the other part will have a CP of 2 (=6-4). Accordingly the stream-2 is split as shown in Figure (a-3).

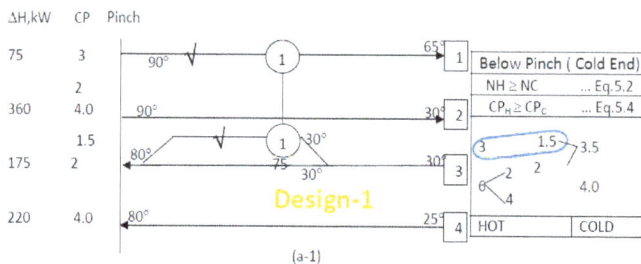

(a-1)

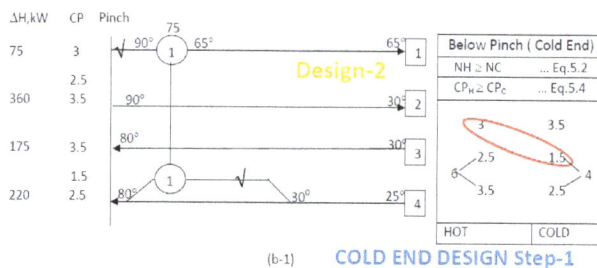

(b-1)   COLD END DESIGN Step-1

The exchanger-2 having load 220 kW is now placed between split part of stream-2 having CP equal to 4 and stream 4 and the exchanger -3 having load 100 kW is placed between split part of stream-3 having CP equal to 2 and the split part of stream-2 having CP equal to 2. All matchings are shown in CP table with blue bold lines. Temperatures after each matching is computed and placed in the HEN to show that no violation of ΔT_{min} occurs during design. The final design for cold end is shown in Figure (a-3).

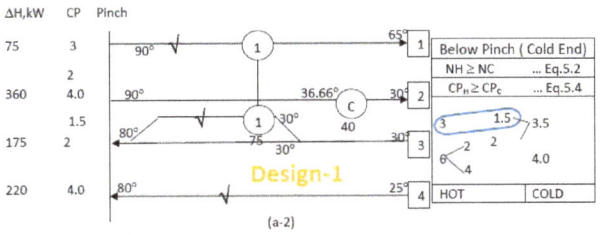

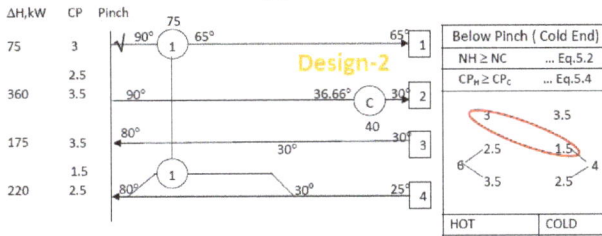

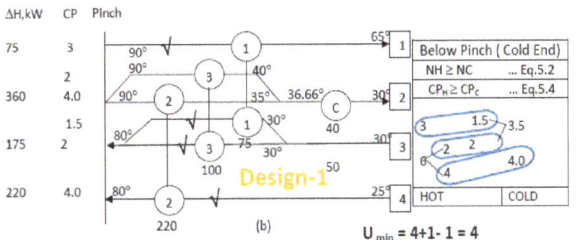

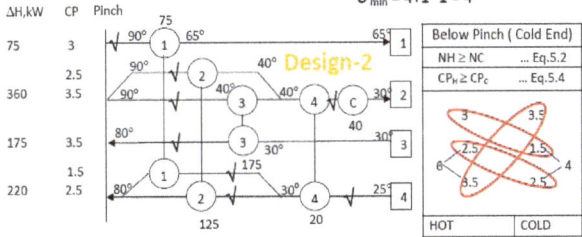

Different steps (a-1) to (a-3) for feasible design-1 and & (b-1) to (b-3) for feasible design-2 of the cold end design for problem in Table.

Design-2

For design-2 the stream No.4 is split in to two parts to offer a possibility for matching of stream- 1 with a part of stream-4. Thus, stream-4 is split in such a way that one part of it will tick off stream-1. Therefore, stream-4 is split into two parts having CPs equal to

1.5 and 2.5. The stream having CP equal to 1.5 will offer a load of 75 kW. Now stream-1 is matched with a part of stream-4 having CP equal to 1.5. This ticked off stream-1 as well as a part of stream-4 having CP equal to 1.5 as can be seen from Figure (b-1). A heat balance shows that the initial temperature of the split stream should be 30°C to offer 75 kW load. Accordingly the initial temperature of this stream is marked as 30°C in the HEN. This matching is shown in CP table by red solid line.

As we have only one hot stream (stream-2) is left, the cooler has to be placed on it. Thus a cooler is placed away from the pinch point on this stream as shown in Figure (b-2). The load of the cooler is 40 kW as reported in the problem $(Q_{C\min} = 40\,kW)$. A simple computation shows that the input temperature to cooler will be 36.66°C.

Now stream-2 is split in two parts to provide balanced heat (after cooler) to satisfy the heat load of split part of stream 4 having CP equal to 2.5 and stream-3 having CP equal to 3.5. As the CP of stream 3 is 3.5, one of the split part of the stream-2 should have a CP equal to 3.5 to match CP criterion. Thus the other part will have a CP of 2.5 (=6-3.5). Accordingly the stream-2 is split as shown in Figure (b-3).

The exchanger-3 having load 175 kW is now placed between split part of stream-2 hav-ing CP equal to 3.5 and stream 3 and the exchanger -2 having load 125 kW is placed be-tween split part of stream-4 having CP equal to 2.5 and the split part of stream-2 having CP equal to 2.5. Finally to maintain heat balance between stream-2 and stream-4, heat exchanger-4 with load 20 units is placed.

All matchings are shown in CP table with red bold lines. Temperatures after each matching is computed and placed in the HEN to show that no violation of $\Delta T_{\min}$ occurs during design. The final design for cold end is shown in Figure (b 3).

Final Design

The final MER design of the HEN is achieved by joining the Hot end design with cold end design. For the present case as two designs are proposed for hot as well as cold end a total of four feasible final design is possible. This is due to the facts that hot end as well as cold end design is in balance as far as load is concerned.

Table : Alternate final designs for HEN

Feasible HEN configurations	Hot End Design	Cold End Design	No. of Units.
1.	Design-1	Design-1	07
2.	Design-2	Design-2	08
3.	Design-1	Design-2	08
4.	Design-2	Design-1	07

A minimum units target applied to the complete problem shows that the number of units should be five (=4+2-1). Thus there is a scope to decrease number of units by two.

Invariably the total number of units in a MER design like the above has more number of units due to the presence of loops in the HEN design. Thus the removal of loops from the design offers a scope to decrease number of units and to reduce the fixed cost. But this decrease is achieved at the cost of increased utility cost as during this process heat flows through pinch point and the design no more remains MER.

The final design for hen configuration 1 is shown in Figure.

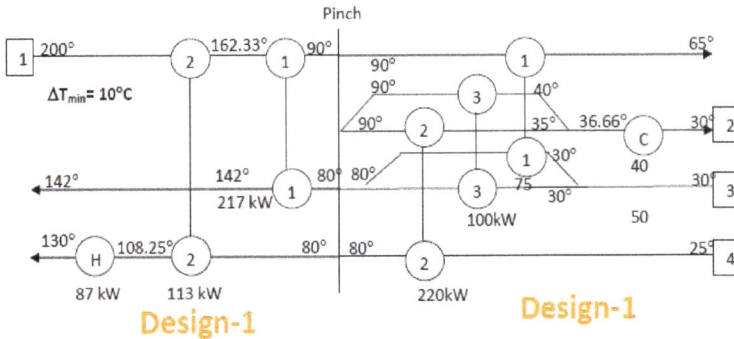

complete design of HEN for problem in Table

After demonstrating the PDM with a sample problem, three design problems will be taken demonstrate design of problems such as Threshold problem, Single pinch and multiple pinch respectively.

The characteristics of these problems are that these either need hot utility or cold utility and not both. Threshold problems can be divided into two broad categories for purpose of design. In the first type, the closest temperature approach between the hot and cold composites is at the "non-utility" end and the curves diverge away from this point. The second type, there is an intermediate near-pinch, which can be identified from the composite curves as a region of close temperature approach. Now the question is whether pinch design method(PDM) should be modified to deal with design of threshold problems.

The logic behind PDM is to start the design where the problem is most constricted. If the design problem has a pinch then the problem is most constricted at pinch and thus it should start from pinch point moving away from it. If the pinch(or most restricted part) is at the "non-utility" end then it should start from there. Let us explain it through as example problem given in Table:

Table: Four stream threshold problem with pseudo pinch for design of HEN ($\Delta T_{min} = 5^0 C$)

Stream No. & Type	CP (kW/ K)	Actual Temperatures (°C)	
		Supply Temp.	Target Temp.
Hot 1	3	180	60
Hot 2	1.5	140	30

| Cold 1 | 2 | 20 | 135 |
| Cold 2 | 4 | 80 | 140 |

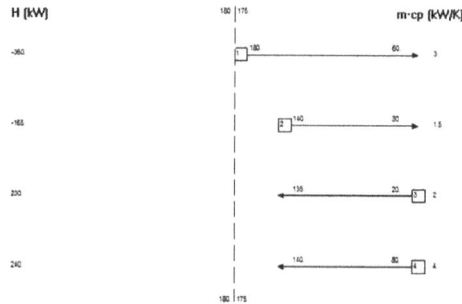

Stream data and pinch for problem shown in Table

Figure shows the stream data and pinch of the problem. Figure shows the hot and cold composite curves for the problem. From this it is clear that the problem requires no hot utility and the pinch point is at hot end. However, the problem exhibits a pseudo pinch as shown in Figure. This is called pseudo pinch as heat (17.5kW) passes through it as evident from Figure.

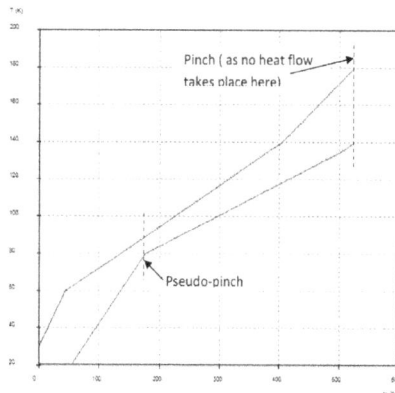

Hot and Cold Composite curves for problem in Table

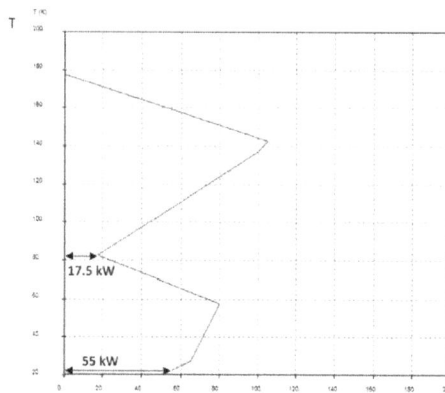

Grand Composite curves for problem in Table

For the present problem hot utility requirement of 0 kW where as the cold utility requirement is 55 kW.

For the present case, the pseudo hot pinch is at 85 °C and that of pseudo cold pinch is at 80 °C. Heat flow through this pseudo pinch is 17.5 kW.

Thus, in this case, it is advisable to treat the problem as a pinch problem and pseudo pinch as pinch and design away from the pseudo pinch. The only difficulty in treating this problem as pinch problem is that one half of the problem (the hot end as shown in Figure) will not offer any flexibility to match against hot utility stream.

It should be noted that the hot end of the problem is not in heat balance. The enthalpy available with hot stream (3*(180-85)+1.5*(140-85) = 367.5 kW) is 367.5 kW where as heat required by cold streams 3&4 is (2.0*(135-80)+4.0*(140-80)= 350 kW). The difference between these two figures is 17.5 kW. This amount will pass to cold end section of the problem.

Hot End Design

The grid diagram of the hot end of the problem is shown in Figure. For the design of the hot end, as per the CP rules, a match of 240 kW between the stream-1 and stream-4 is placed which ticks off stream-4. Now Exchanger-2, having a capacity of 82.5 kW is placed between stream-2 and stream-3 ticking off stream-2. Further, stream 3 has a capacity to take 110 kW of it of which 82.5 kW has been passed by exchanger-2. The rest, 27.5 kW, can be passed through exchanger-3 placed between stream-1 and stream-3 ticking off stream-3. Now the only stream which has residual heat (17.5 kW) left is stream-1. Thus stream-1 will not reach the pseudo point of 85°C, instead it will reach to 90.83°C so that heat equal to 17.5 kW is passed to cold end of the problem. Please note that the earlier pseudo pinch temperature of stream-1 was assumed to be 85°C. Now the pseudo pinch temperature of stream-1 being 90.83°C the stream is ticked off. The complete design of the hot end of the problem is shown in Figure. The temperature levels on the hot end part of the grid diagram clearly shows that there is no violation of ΔT_{min} in this part of HEN.

Cold End Design

Cold end design is shown in Figure. In this part of the problem, heat available with stream -1 is 92.5 kW(=17.5+3*(85-60)=3*(90.83-60)=92.5kW) and that of hot stream -2 is 82.5 kW(1.5*(85-30)) making a total heat available to be 175 kW. The cold stream-3 can only take up (2*(80- 20)=120 kW) 120 kW heat. Hence the rest heat, 55 kW will be taken up by the cold utility as computed for the problem. To start the the design exchanger-4 having a load of 92.5 kW is placed between stream-1 and stream-3 ticking off the stream-1. Then exchanger -5, having capacity of 27.5 kW is placed between stream-2 and 3 ticking off stream-3. The excessive heat load of stream-2 which is 55 kW is transferred to the cooler.

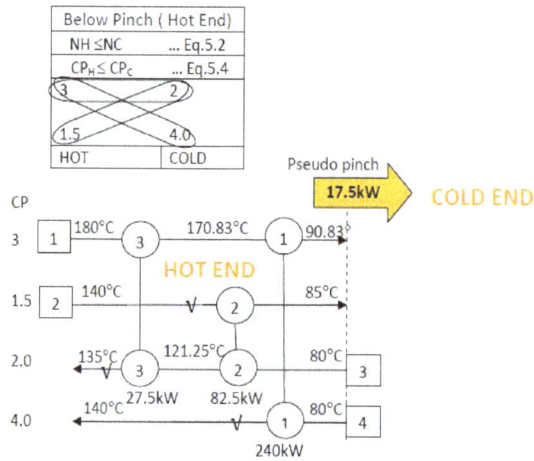

Hot end HEN design for threshold problem treating it as pinched problem

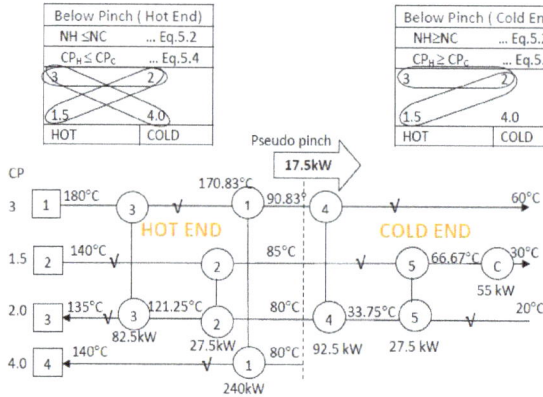

Cold end HEN design for threshold problem treating it as pinched problem

The second problem as shown in Table is considered to shows a different type of threshold problem. The hot and cold composite curves are shown in Figure (a) and GCC is shown in Figure (b). From the composite curves it is clear that the problem requires no hot utility. However, cold utility of amount 239.5 kW is required. The most constrained part of this problem is the non-utility end where temperature differences are constrained. Such problems are treated as one half of the pinch problem and design is strated where the problem is most constrained, i.e. non –utility end moving way from it. For the present case one feasible HEN designs is developed as shown in Figure.

Table: Four stream problem utility prediction for ΔT_{min} equal to 10°C.

Name of the stream	Supply Temperature Ts, °C	Target Temperature Tt, °C	CP kW/°C	ΔH kW
Hot-1	190	55	3.5	-472.5
Hot-2	155	40	1.8	-207
Cold-1	20	140	2	240
Cold-2	70	150	2.5	200

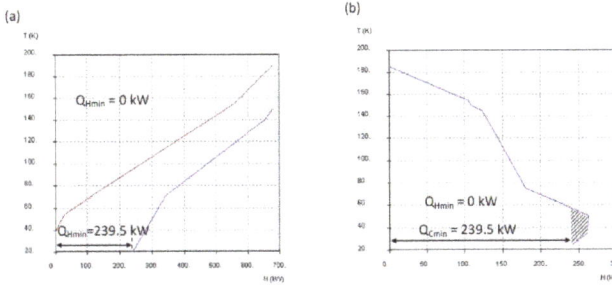

Composite curves and GCC for threshold problem of Table

Design

For cold end design number of hot stream should be greater or equal to number of cold streams. In the present case there are two hot streams and two cold streams, thus it satisfies the above rule. From the observation it is clear that CP of stream -2 is 1.8 and that of stream 3 and 4 are 2.0 and 2.5 respectively. Thus no direct pinch match can be placed between stream-2 and stream-3 as it will violate CP rule. Thus it was decided to split the stream-3 two part having CP values 1.725 and other 0.275. This will upset the rule $NH \geq NC$. Thus to avoid it stream-1 is split in to two parts having CP values 3.004 and 0.496.

Now exchanger-1 having load 200kW is put between split part of stream-1(CP=3.004) and stream-4(CP=2.5) to tick off stream-4. This satisfies the CP rule $CP_H \geq CP_C$. The exchanger-2 having load equal to 207 kW is placed between split part of stream-2 (CP=1.725) and stream-2 ticking off stream-2. The exchanger-3 having load equal to 33 kW is put between the split stream-1 (CP=0.496) and split part of stream-2(CP=0.275). Having placed exchanger-2 and 3, now the stream-2 is completely ticked off. Finally a cooler having capacity of 239.5 kW($= Q_{Cmin}$) is placed on stream -1 to complete the heat balance. The complete design is shown in Figure.

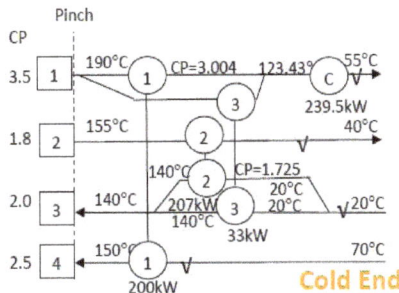

Feasible HEN designs for threshold problem in Table

To demonstrate the design of single pinch problem, problem given in Table is considered. For this purpose a fairly complex eight stream problem is taken up.

Table: Eight stream MER design of HEN for $\Delta T_{min} = 10^0 C$

Stream Name & Type	CP (kW/ K)	Actual Temperatures (^{0}C)	
		Supply Temp.	Target Temp.
H-1	0.1	150	40
H-2	0.15	140	30
H-3	0.15	130	25
H-4	0.2	150	30
C-1	0.1	20	140
C-2	0.15	15	130
C-3	0.25	25	145
C-4	0.3	80	140

For the present problem, hot pinch temperature is 90 °C and Cold pinch temperature is 80 °C. Further, hot Utility requirement is 16.25 kW whereas, the cold utility requirement is 6.25 kW.

The grid diagram, composite curve and the grand composite curve for the above problem are shown in Figures respectively.

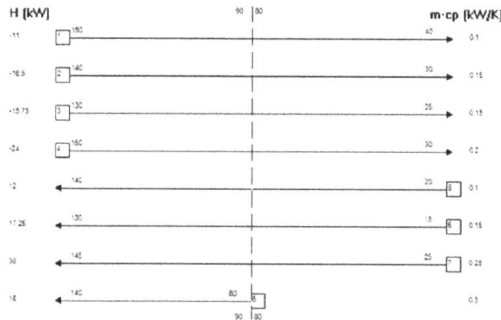

Stream data and pinch for problem shown in Table

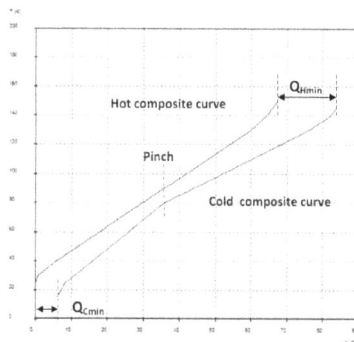

Hot and Cold composite curves for problem in Table

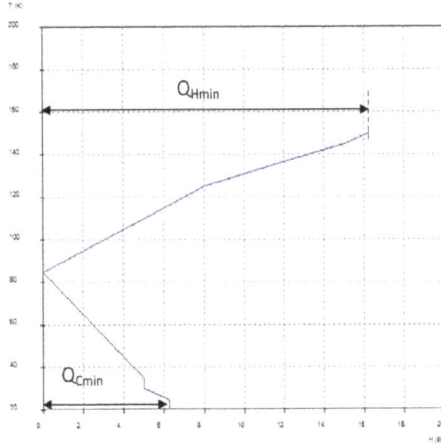

GCC for problem in Table

Hot End Design

The hot end design of the problem given in Table is shown in Figure. The CP table with CP rules and number stream rules are also given to guide the design. As CP of the stream -1 is equal to stream-5 a match (exchanger-1) can be placed between these. The load of the match is 6 kW. This ticks off stream-5 & 1. The second match, exchanger-2 with load 7.5 kW is placed between stream -2 and stream-6 as per CP rules. This ticks off stream-6 & 2. The third match, exchanger-3 having a load 6 kW is placed between stream-3 and stream-8 as per CP rule. This ticks off stream-3. The fourth match, exchanger-4 with 12 kW load is placed between stream-4 and stream-7 ticking off stream-4. After the above said matching only the cold streams 7 & 8 needs heating of 4.25 kW and12 kW respectively which was supplied through heater. This completes the hot end design of the problem as shown in Figure. It should be noted that number of units in this design is two less that the targeted value of 8. This is due to the presence of two subset equality in the problem.

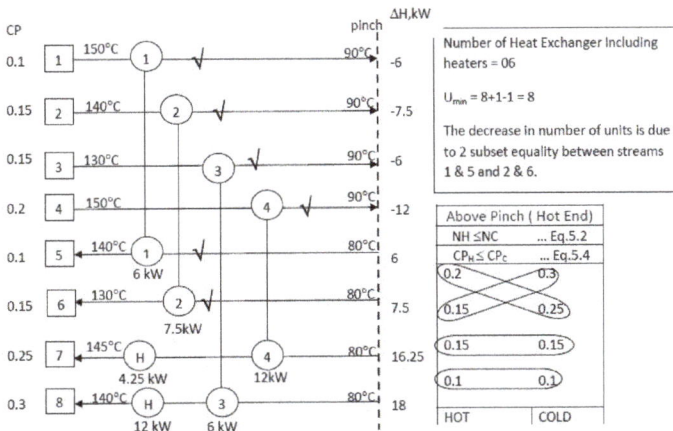

MER design of Hot end of HEN for single pinch problem

Cold End Design

The cold end design of the problem given in Table is shown in Figure. The CP table with CP rules and number stream rules are also provided to guide the design. As CP of the stream -2 is greater than CP of stream-5. Hence, a match (exchanger-5) is placed between these. The load of the match is 6 kW. This ticks off stream-5. The second match, exchanger-6 with load 9.75 kW is placed between stream -3 and stream-6 as per CP rules. This ticks off stream-6 & 3. The third match, exchanger-7 having a load of 8.75 kW is placed between stream-4 and split stream of stream-8 having CP equal to 0.2 as per CP rule. The fourth match, exchanger-8 with 5 kW load is placed between stream-1 and stream-7 ticking off stream-1 & 7. It should be noted that this match is not as per CP rules. However, this does not violet the ΔT_{min} criterion as the exchanger- 8 is not a pinch exchanger. After the above said matching only the hot streams 2 & 4 needs cooling of 3 kW and 3.25 kW respectively which was supplied through coolers. This completes the cold end design of the problem as shown in Figure. It should be noted that number of units in this design is one less that the targeted value of 7. This is due to the presence of one subset equality in the problem.

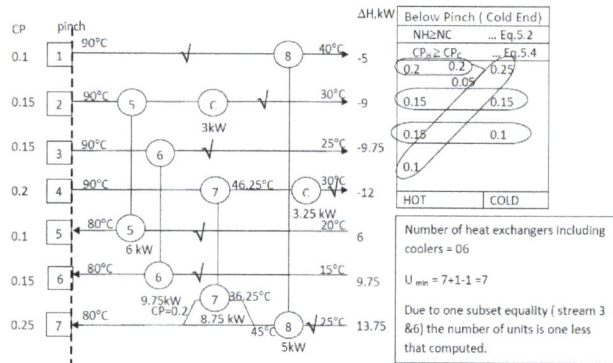

MER design of Cold end of HEN for single pinch problem

Complete Design of HEN

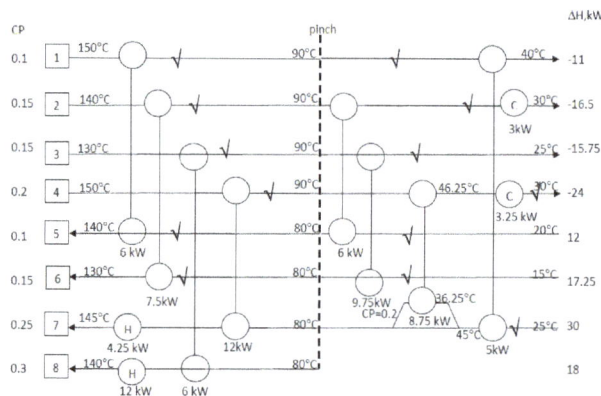

Grid diagram of MER design of the HEN for problem in Table

By joining the hot end design and cold end design one can achieve the complete design of the HEN for the problem. This is done by joining the hot end design of the HEN shown in Figure and cold end design shown in figures. This the complete design is available in Figure as shown below.

To demonstrate the design of multiple pinch problem the problem shown in Table is considered. The hot utility and cold utility demand for the process is 10 kW and 14.7 kW respectively. The hot process pinch temperature is 160 °C and that of cold process pinch temperature is 150 °C.

Table: A four stream problem to demonstrate design of multiple pinch problem ($\Delta T_{min} = 10^0 C$)

Stream No. & Type	CP (kW/ K)	Actual Temperatures (°C)	
		Supply Temp.	Target Temp.
Hot 1	0.1	270	35
Hot 2	0.22	220	60
Cold 1	0.18	40	190
Cold 2	0.3	150	240

The hot and cold composite curves for the problem are presented in Figure. The GCC of the problem, Figure, shows that two hot utilities namely high pressure steam at 252 °C and low pressure steam at 192.85 °C can be used to satisfy the hot utility demand of 10 kW. The actual consumption of high pressure steam as hot utility is 4.75 kW and that of low pressure steam is 5.25 kW. Once, these hot utilities are used in the process, the low pressure steam creates an utility pinch in the process. Thus we have two pinches now, one process pinch and the other utility pinch as shown in Figure in the process. Based on the above modification, the problem shown in Table is reformulated as a multiple pinch problem as given in Table.

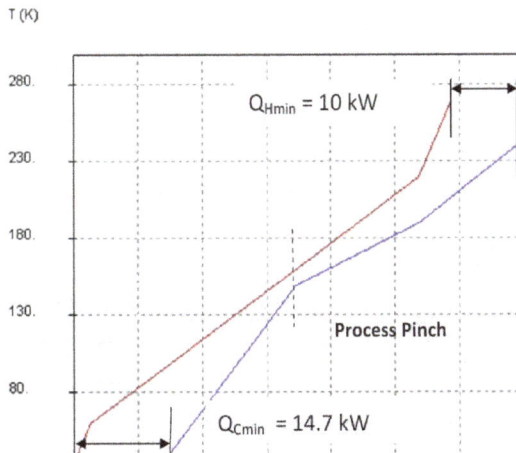

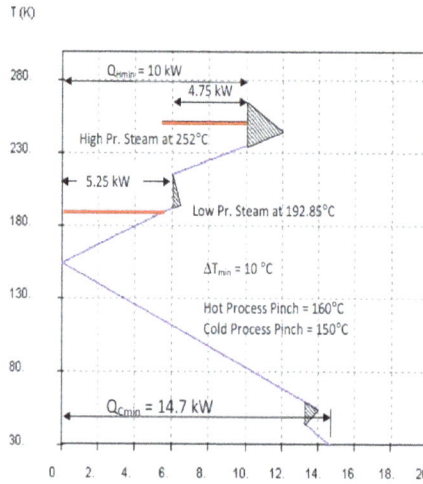

GCC and placement of multiple hot utility

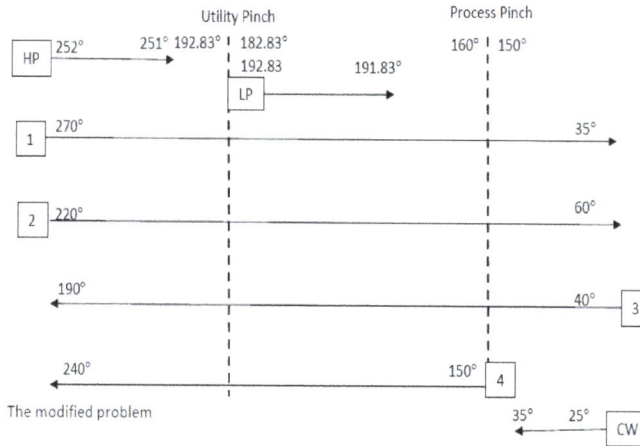

Stream data for the modified problem having process as well as utility pinches

Table: Modified problem having multiple pinch (for $\Delta T_{min} = 10^0 C$)

Stream No. & Type	CP (kW/ K)	Actual Temperatures (^{0}C)	
		Supply Temp.	Target Temp.
Hot 1	0.1	270	35
Hot 2	0.22	220	60
Hot-3(High Pr. Steam)	4.75	252	251
Hot-4(Low Pr. Steam)	5.25	192.85	191.85
Cold 1	0.18	40	190
Cold 2	0.3	150	240
Cold-3(Cold water)	1.47	25	35

The stream data along with the position of process as well as utility pinch is shown in Figure. The process pinch is at 160 °C/150 °C and the utility pinch is at 192.83 °C/182.83 °C. The Balanced composite curve with high and low pressure stream and cooling water profiles is shown in Figure. The positions of process as well as utility pinches are also marked in this figure.

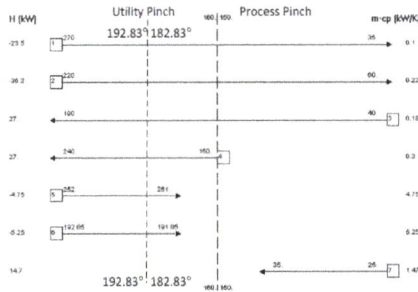

Stream data and placement for the modified problem

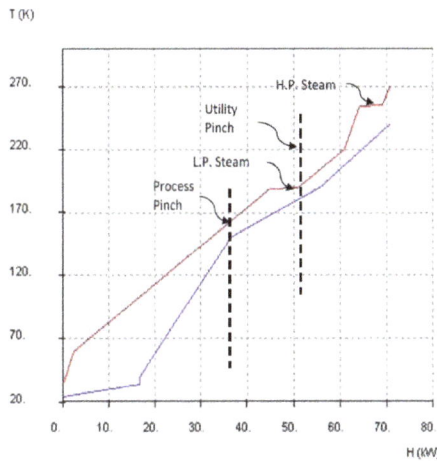

Balanced composite curve(BCC) for Problem in Table

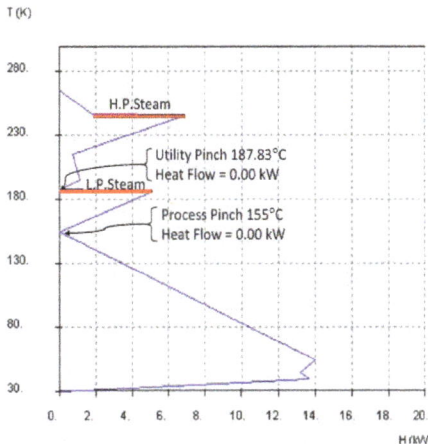

Grand comosite curve for problem in table

The grand composite curve (GCC) for the problem is shown in Figure with position of High Pressure steam and low pressure steam. Please not that GCC is plotted for shifted temperatures and not actual temperatures and only due to this the pinch temperature reported is 155 °C. The actual hot pinch temperature is $160^0 C(=155^0 C + \Delta T_{min} / 2 = 160^0 C)$ and that of cold pinch temperature is 150 °C. Similarly the reported shifted utility pinch temperature is 187.83 °C.

Figure shows the grid diagram when two steam levels are used with the utility pinch dividing the process into three parts. The same is shown schematically in Figure along with direction of movement of design process.

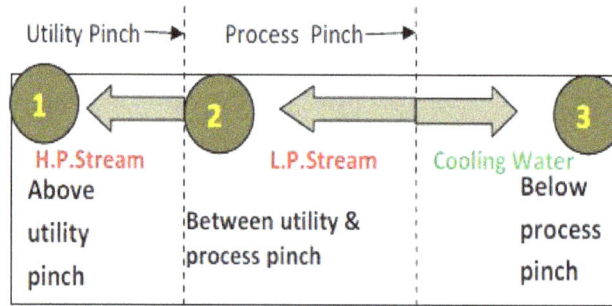

Conceptual design direction for the problem

As per the pinch rules, heat should not be transferred across either through the process pinch or the utility pinch by process-to-process heat exchange. Further, there should not be inappropriate use of utilities meaning that above the utility pinch, high-pressure steam should only be used and no low-pressure steam or cooling water be used. Between the utility pinch and the process pinch only low-pressure steam should be used and no high-pressure steam or cooling water be used. Further, below the process pinch only cooling water should be used. These appropriate utility streams have been inducted with the process streams in Figure.

The network for the above problem can now be designed using the pinch design method. The pinch design method(PDM) starts the design at the pinch, and moves away. For pinch matches, the rules for the CP inequality and the number of streams must be complied. Above the utility pinch and below the process pinch as shown in Figure, there is no problem in applying PDM. However, between utility and process pinches there is some confusion, since designing away from both pinches would lead to a conflict.

However, a close examination of the part between two pinches reveals that, one is more constrained than the other. Below the utility pinch, the CP rule, $CP_H \geq CP_C$ is required and low- pressure steam is available as a hot stream with an large CP. Therefore, the above rule can be satisfied easily. Thus, observing the rules of PDM, which tells that the design should start from the most constrained pinch, the present design, for the part between pinches, should start from process pinch and move away as shown in Figure.

Once, the direction of movement of design procedure is established as given in Figure, the process of design is divided into three parts as given below:

1. Design of HEN above the utility pinch (Hot End design)

2. Design of HEN for section between utility pinch and process pinch

3. Design of HEN below the process pinch.

Design of HEN above the Utility Pinch (Hot End Design)

The design of HEN is shown in Figure. It starts from utility pinch and moves away from it. In this portion of the HEN, the high pressure steam is only used. It should be noted that in this part of the problem, the number of process hot stream is 2 and that of process cold stream is 2. Thus to find a match for the H.P. Steam the stream-4 having a high value of CP equal to 4 is split into two parts having CPs equal to 0.22 and 0.08. To start with, exchanger-1 with load 1.2906 kW is placed between stream-1 and stream-3 (CP of Stream-3 $\geq$ CP of Stream-1). As per CP rules, the second match, exchanger-2 with a load 5.977 kW is placed between stream-2 and split part of stream-4 having CP equal to 0.22. Obeying the CP rule the third match, exchanger-3 with load 2.1736 kW, is placed between stream-1 and split part of stream-4 having CP equal to 0.08. The exchanger-4 having load equal to 4.253 kW is placed between steam-1 and 4. The, last exchanger-5 having load equal to 4.75 kW $(= Q_{H\,min})$ is placed between high pressure stream(H.P. Steam) and stream -4 having CP equal to 0.3. A check shows that there is no violation of ΔT_{min} criterion and thus the design is accepted.

The complete design of the HEN for the part above the utility pinch is shown in Figure where the design started at utility pinch and moved away from it as shown in Figure.

Design above Utility Pinch (Hot End)

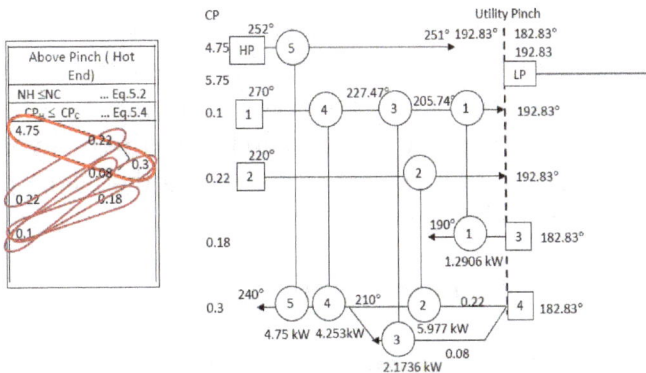

HEN design above the utility pinch (hot end)

Design of HEN between Utility Pinch and Process Pinch

The design of HEN for this part of the problem is shown in Figure. It starts from process pinch and moves away towards utility pinch. In this portion of the HEN, the low pressure steam is only used. This part is treated as hot end design with respect to process pinch as the design moves away from process pinch towards utility pinch. It should be noted that in this part of the problem, the number of process hot stream is 2 and that of process cold stream is 2. To start with, exchanger-6 with load 3.283 kW is placed between stream-1 and stream-3 (CP of Stream-3 ≥ CP of Stream-1). In accordance with the CP rules, the second match, exchanger-7 with a load 7.226 kW is placed between stream-2 and stream-4. It should be noted that splitting of L.P. Stream is not mandatory. However, to satisfy the need of stream-3 & 4 two.

HEN Optimization

In MER HENs cold and hot utilities as exactly same as targeted. However, MER HENs are not the best HENs as far as total annual cost(TAC) is concerned. Now the question is whether, the MER HENs designed ? The answer to this question is yes.

Using the PDM a HEN structure can be created based on the assumption that no heat exchanger should have a temperature difference less than ΔT_{min}. However, the structure thus created can't be termed optimum and some procedure for optimization can be adopted. Let us one by one see the drawbacks of the HEN created by PDM.

In an MER design, the pinch divides the problem into two thermodynamically independent regions as shown in Figure (a). HEN designs are performed independently for these two parts and once done are joined together to get a complete HEN for the problem. Due to this process some weaknesses creep into the design.

Now, suppose the design is converted in to a non-MER design by allowing β amount of heat to pass through pinch division, then the hot and cold utility amount will rise to $Q_{H\,min} + \beta$ and $Q_{C\,min} + \beta$ respectively as shown in Figure (b). Now the earlier regions(Hot end & cold end) defined in MER design is no longer thermodynamically independent. Thus, if the number of units target is applied to the whole problem, ignoring pinch division, then one will observe that:

$$U_{min}(Non-MER) \le U_{min}(MER)$$

The above result is an expected one and can be explained as follows: When one applies unit target to a MER design, streams that cross pinch division are counted twice and thus in a MER design the total number of units are always more than a non-MER design. This clearly shows that by allowing heat to transfer through pinch division, though the amount of utility increases, the number of units is decreased and thus offers an opportunity to decrease the fixed cost of the HEN. Thus there is a trade-off between energy recovery and number of units employed.

Trading off units and Energy

The above facts are explained through the example given in Table.

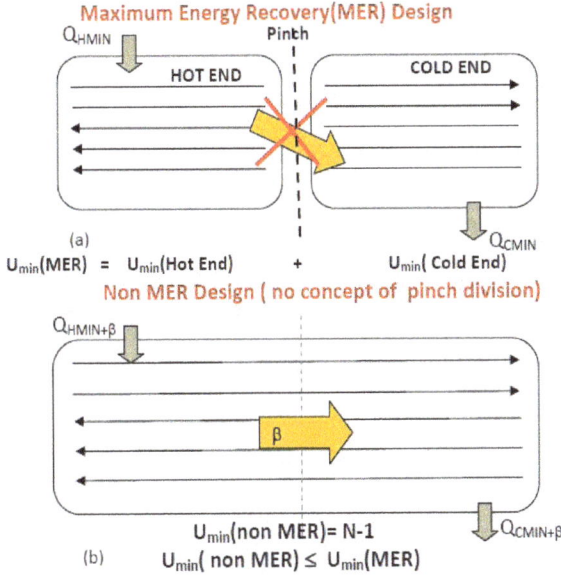

Targeting for number of units in (a) MER as well as in (b) non-MER designs

Table: A four stream problem to demonstrate energy vs. no. of units trade off ($\Delta T_{min} = 10^0 C$)

Stream Name	Supply Temperature	Target Temperature	CP
H-1	160	60	2.5
H-2	140	30	1.5
C-1	20	125	2
C-2	80	130	3.5

The hot pinch temperature for the above problem is 900 C whereas, the cold pinch temperature 800 C. The hot utility requirement is 15 kW whereas that of cold utility requirement is 45 kW. The stream diagram is shown in Figure. The hot end as well cold end design of the HEN was carried out for the problem. After joining the hot and cold end HENs the overall MER HEN is shown in Figure. It can be seen that the overall number of units in this design is 6 including heater and coolers whereas a non-MER design should produce 5 (=4+2-1) units. This clearly shows that there is a scope of decreasing one unit in a non-MER design however, at the cost of increased hot and cold utility which is to be investigated to strike a trade-off. The reason for one extra number of unit is due to the presence of a loop in the MER design. Thus if the loop is broken then one unit can be reduced in the design. However, this will convert the MER design to a non-MER design and some amount of heat will pass through pinch division. Now the question is what should be the proper way to do it so that we get the benefit of reduced number of units in the HEN and at the same time the amount of heat passed through the pinch is also not much.

Figure shows an attempt to break the loop by removing the heat exchanger (45 kW) placed at one side of the loop and adding this load with the heat exchanger (75kW) on the other side of the loop making its load as 120 kW. From common sense it appears that this arrangement will work. However, it violates the ΔT_{min} criterion and thus the design is not acceptable. Figure shows an alternate arrangement where the amount of hot as well as cold utilities are increased by 45kW (load of the removed heater). Theoretically, this design works and does not violates the ΔT_{min} criterion and thus acceptable. However, the utility loads are increased significantly.

Now the question is can there be a third alternative to the design which will decrease the utility loads but will provide a feasible design? The answer is yes, it can be done by utilizing the utility path as discussed below:

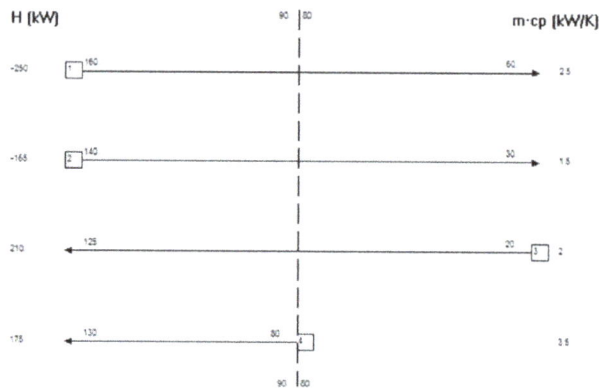

Stream diagram for the problem shown in Table

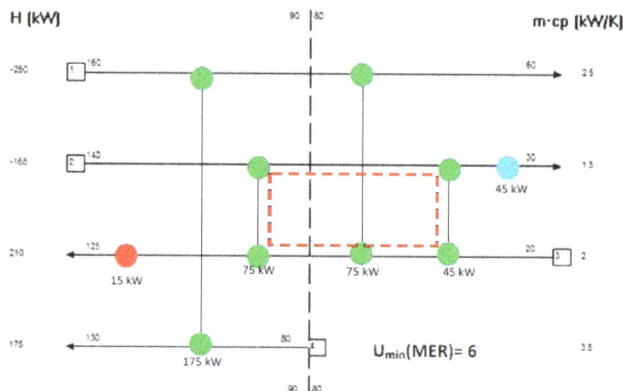

Grid diagram of the HEN (MER Design)

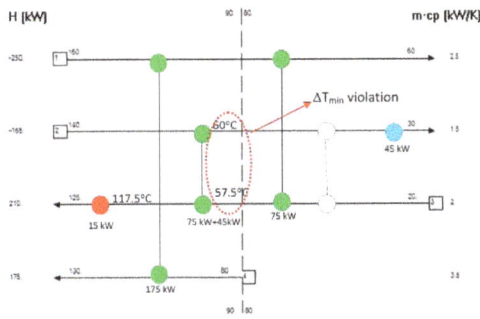

Grid diagram of the HEN (MER Design)

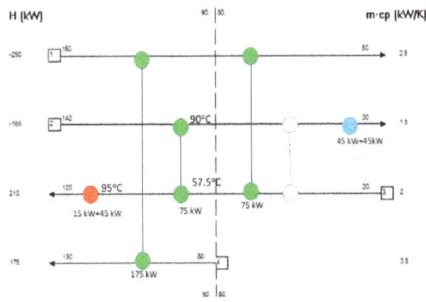

Grid diagram of the HEN (Non-MER Design)

Given the HEN as in Figure it is possible to identify loops and utility paths. In the Figure a loop which is shown by red dotted lines connects exchangers B and D. It is possible to trace a close path in the network for a loop. Though the existence of a loop introduces some element of flexibility by transferring load from one unit to other in the loop, it adds one more unit per loop. For example, the load of exchanger "B' can be transferred to "D" which is present in the same loop. Further, by making the load of an exchanger zero the exchanger can be removed from the loop and the loop can be broken. For example if the load of exchanger "D" is made zero and to compensate the load the load of exchanger "B" is increased to 120 (=75+45) kW, then the loop -1 can be broken. However, this will disturb the heat flow in the network and violation of ΔT_{min} will take place which has to be restored to keep the utilities near targeted value.

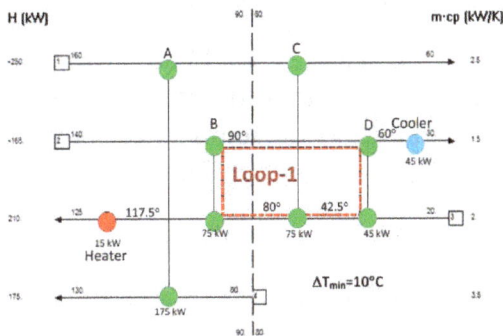

Identification of loop(red dotted line) in the MER HEN design

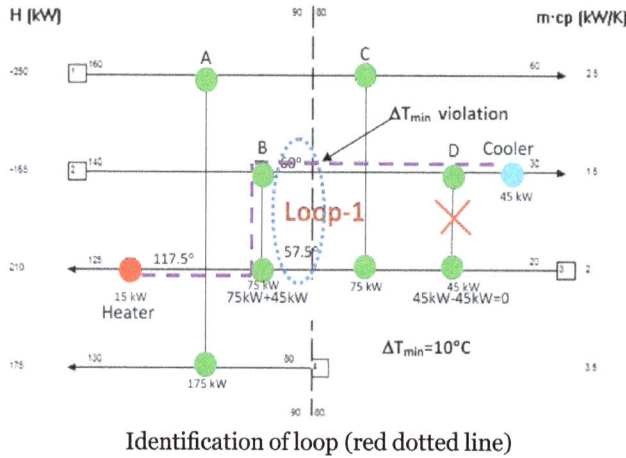

Identification of loop (red dotted line)

The restoration of the ΔT_{min} can be done through a utility path. Utility paths are those paths which connects two different utilities. This path could be path joining hot utility, steam to cold utility cold water or a path from high pressure steam to low pressure steam.

For the present case in Figure a utility path from heater through exchanger "B" to cooler(purple dotted line) can be traced to find a ΔT_{min} compensation plan for the problem.

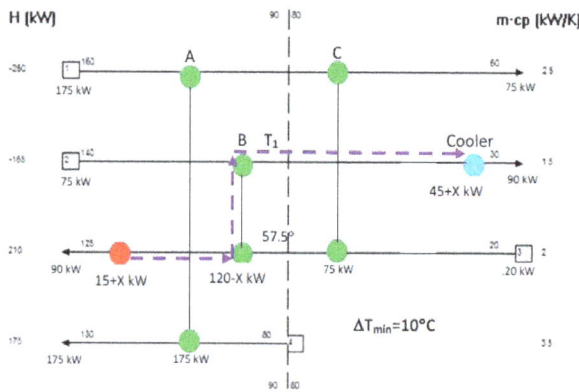

Identification of loop(red dotted line) in the MER HEN design

If we add heat load of "X" to the heater, it load becomes 15+X kW, then for enthalpy balance the load of exchanger "B" should be decreased by "X" amount and the load on cooler be increased by "X" amount. Let the temperature of stream-2 be T_1 after the exchanger "B". Now to restore the ΔT_{min}, T_1-57.5° should be equal to 10 °C.

The above is possible when following equation is satisfied.

$140° - (120-X)/1.5 = T1 = 57.5°+10°=67.5$ °C

Or $30 + (45+X)/1.5 = T1 = = 57.5°+10°=67.5$ °C

To satisfy the above equation the value of "X" should be 11.25 kW. The above solution is clearly better than the solution offered in Figure.

PDM creates network structures based on the assumption that none of the exchanger in the network should have a temperature difference less than ΔT_{min}. Once a HEN structure is designed based on PDM, it can be subjected to optimization to improve its cost effectiveness. The constraint imposed on it can be removed one by one to get improved solution in terms of fixed cost of HEN, number of units or shells in HEN or finally TAC of HEN.

1. The constraint that none of the exchanger should have temperature difference less than ΔT_{min} can be relaxed

2. The exchanger duties could be redistributed or even set to zero to remove one of it.

3. Removal of extra number of units in the HEN by breaking loops.

4. Proper selection of utility paths in the HEN after loop breaking to restore ΔT_{min}. Utility paths and loops provide degree of freedom for optimization and should be utilized to improve cost effectiveness of HEHs.

5. Stream splitting also offers an additional degree of freedom in the process of optimization and should be manipulated for the improvement of HENs.

Mathematical Modeling in Evaporator and Distillation

A mathematical model uses concepts of mathematics. These models can be used in various subjects like physics, biology, psychology, sociology and artificial intelligence. The chapter closely examines the key concepts of mathematical modeling to provide an extensive understanding of the subject.

Mathematical Model

A mathematical model is a description of a system using mathematical concepts and language. The process of developing a mathematical model is termed mathematical modeling. Mathematical models are used in the natural sciences (such as physics, biology, earth science, meteorology) and engineering disciplines (such as computer science, artificial intelligence), as well as in the social sciences (such as economics, psychology, sociology, political science). Physicists, engineers, statisticians, operations research analysts, and economists use mathematical models most extensively. A model may help to explain a system and to study the effects of different components, and to make predictions about behaviour.

Elements of a Mathematical Model

Mathematical models can take many forms, including dynamical systems, statistical models, differential equations, or game theoretic models. These and other types of models can overlap, with a given model involving a variety of abstract structures. In general, mathematical models may include logical models. In many cases, the quality of a scientific field depends on how well the mathematical models developed on the theoretical side agree with results of repeatable experiments. Lack of agreement between theoretical mathematical models and experimental measurements often leads to important advances as better theories are developed.

In the physical sciences, the traditional mathematical model contains four major elements. These are

1. Governing equations
2. Defining equations

3. Constitutive equations

4. Constraints

Classifications

Mathematical models are usually composed of relationships and *variables*. Relationships can be described by *operators*, such as algebraic operators, functions, differential operators, etc. Variables are abstractions of system parameters of interest, that can be quantified. Several classification criteria can be used for mathematical models according to their structure:

- Linear vs. nonlinear: If all the operators in a mathematical model exhibit linearity, the resulting mathematical model is defined as linear. A model is considered to be nonlinear otherwise. The definition of linearity and nonlinearity is dependent on context, and linear models may have nonlinear expressions in them. For example, in a statistical linear model, it is assumed that a relationship is linear in the parameters, but it may be nonlinear in the predictor variables. Similarly, a differential equation is said to be linear if it can be written with linear differential operators, but it can still have nonlinear expressions in it. In a mathematical programming model, if the objective functions and constraints are represented entirely by linear equations, then the model is regarded as a linear model. If one or more of the objective functions or constraints are represented with a nonlinear equation, then the model is known as a nonlinear model. Nonlinearity, even in fairly simple systems, is often associated with phenomena such as chaos and irreversibility. Although there are exceptions, nonlinear systems and models tend to be more difficult to study than linear ones. A common approach to nonlinear problems is linearization, but this can be problematic if one is trying to study aspects such as irreversibility, which are strongly tied to nonlinearity.

- Static vs. dynamic: A *dynamic* model accounts for time-dependent changes in the state of the system, while a *static* (or steady-state) model calculates the system in equilibrium, and thus is time-invariant. Dynamic models typically are represented by differential equations or difference equations.

- Explicit vs. implicit: If all of the input parameters of the overall model are known, and the output parameters can be calculated by a finite series of computations, the model is said to be *explicit*. But sometimes it is the *output* parameters which are known, and the corresponding inputs must be solved for by an iterative procedure, such as Newton's method (if the model is linear) or Broyden's method (if non-linear). In such a case the model is said to be *implicit*. For example, a jet engine's physical properties such as turbine and nozzle throat areas can be explicitly calculated given a design thermodynamic cycle (air and fuel flow rates, pressures, and temperatures) at a specific flight condition and power setting, but the engine's operating cycles at other flight conditions and power settings cannot be explicitly calculated from the constant physical properties.

- Discrete vs. continuous: A discrete model treats objects as discrete, such as the particles in a molecular model or the states in a statistical model; while a continuous model represents the objects in a continuous manner, such as the velocity field of fluid in pipe flows, temperatures and stresses in a solid, and electric field that applies continuously over the entire model due to a point charge.

- Deterministic vs. probabilistic (stochastic): A deterministic model is one in which every set of variable states is uniquely determined by parameters in the model and by sets of previous states of these variables; therefore, a deterministic model always performs the same way for a given set of initial conditions. Conversely, in a stochastic model—usually called a "statistical model"—randomness is present, and variable states are not described by unique values, but rather by probability distributions.

- Deductive, inductive, or floating: A deductive model is a logical structure based on a theory. An inductive model arises from empirical findings and generalization from them. The floating model rests on neither theory nor observation, but is merely the invocation of expected structure. Application of mathematics in social sciences outside of economics has been criticized for unfounded models. Application of catastrophe theory in science has been characterized as a floating model.

Significance in the Natural Sciences

Mathematical models are of great importance in the natural sciences, particularly in physics. Physical theories are almost invariably expressed using mathematical models.

Throughout history, more and more accurate mathematical models have been developed. Newton's laws accurately describe many everyday phenomena, but at certain limits relativity theory and quantum mechanics must be used; even these do not apply to all situations and need further refinement. It is possible to obtain the less accurate models in appropriate limits, for example relativistic mechanics reduces to Newtonian mechanics at speeds much less than the speed of light. Quantum mechanics reduces to classical physics when the quantum numbers are high. For example, the de Broglie wavelength of a tennis ball is insignificantly small, so classical physics is a good approximation to use in this case.

It is common to use idealized models in physics to simplify things. Massless ropes, point particles, ideal gases and the particle in a box are among the many simplified models used in physics. The laws of physics are represented with simple equations such as Newton's laws, Maxwell's equations and the Schrödinger equation. These laws are such as a basis for making mathematical models of real situations. Many real situations are very complex and thus modeled approximate on a computer, a model that is computationally feasible to compute is made from the basic laws or from approximate models made from the basic laws. For example, molecules can be modeled by molecular orbital models that are approximate solutions to the Schrödinger equation. In engineering, physics models are often made by mathematical methods such as finite element analysis.

Different mathematical models use different geometries that are not necessarily accurate descriptions of the geometry of the universe. Euclidean geometry is much used in classical physics, while special relativity and general relativity are examples of theories that use geometries which are not Euclidean.

Some Applications

Since prehistorical times simple models such as maps and diagrams have been used.

Often when engineers analyze a system to be controlled or optimized, they use a mathematical model. In analysis, engineers can build a descriptive model of the system as a hypothesis of how the system could work, or try to estimate how an unforeseeable event could affect the system. Similarly, in control of a system, engineers can try out different control approaches in simulations.

A mathematical model usually describes a system by a set of variables and a set of equations that establish relationships between the variables. Variables may be of many types; real or integer numbers, boolean values or strings, for example. The variables represent some properties of the system, for example, measured system outputs often in the form of signals, timing data, counters, and event occurrence (yes/no). The actual model is the set of functions that describe the relations between the different variables.

Building Blocks

In business and engineering, mathematical models may be used to maximize a certain output. The system under consideration will require certain inputs. The system relating inputs to outputs depends on other variables too: decision variables, state variables, exogenous variables, and random variables.

Decision variables are sometimes known as independent variables. Exogenous variables are sometimes known as parameters or constants. The variables are not independent of each other as the state variables are dependent on the decision, input, random, and exogenous variables. Furthermore, the output variables are dependent on the state of the system (represented by the state variables).

Objectives and constraints of the system and its users can be represented as functions of the output variables or state variables. The objective functions will depend on the perspective of the model's user. Depending on the context, an objective function is also known as an *index of performance,* as it is some measure of interest to the user. Although there is no limit to the number of objective functions and constraints a model can have, using or optimizing the model becomes more involved (computationally) as the number increases.

For example, economists often apply linear algebra when using input-output models. Complicated mathematical models that have many variables may be consolidated by use of vectors where one symbol represents several variables.

A Priori Information

To analyse something with a typical "black box approach", only the behavior of the stimulus/response will be accounted for, to infer the (unknown) *box*. The usual representation of this *black box system* is a data flow diagram centered in the box

Mathematical modeling problems are often classified into black box or white box models, according to how much a priori information on the system is available. A black-box model is a system of which there is no a priori information available. A white-box model (also called glass box or clear box) is a system where all necessary information is available. Practically all systems are somewhere between the black-box and white-box models, so this concept is useful only as an intuitive guide for deciding which approach to take.

Usually it is preferable to use as much a priori information as possible to make the model more accurate. Therefore, the white-box models are usually considered easier, because if you have used the information correctly, then the model will behave correctly. Often the a priori information comes in forms of knowing the type of functions relating different variables. For example, if we make a model of how a medicine works in a human system, we know that usually the amount of medicine in the blood is an exponentially decaying function. But we are still left with several unknown parameters; how rapidly does the medicine amount decay, and what is the initial amount of medicine in blood? This example is therefore not a completely white-box model. These parameters have to be estimated through some means before one can use the model.

In black-box models one tries to estimate both the functional form of relations between variables and the numerical parameters in those functions. Using a priori information we could end up, for example, with a set of functions that probably could describe the system adequately. If there is no a priori information we would try to use functions as general as possible to cover all different models. An often used approach for black-box models are neural networks which usually do not make assumptions about incoming data. Alternatively the NARMAX (Nonlinear AutoRegressive Moving Average model with eXogenous inputs) algorithms which were developed as part of nonlinear system identification can be used to select the model terms, determine the model structure, and estimate the unknown parameters in the presence of correlated and nonlinear noise. The advantage of NARMAX models compared to neural networks is that NARMAX produces models that can be written down and related to the underlying process, whereas neural networks produce an approximation that is opaque.

Subjective Information

Sometimes it is useful to incorporate subjective information into a mathematical model. This can be done based on intuition, experience, or expert opinion, or based on convenience of mathematical form. Bayesian statistics provides a theoretical framework for incorporating such subjectivity into a rigorous analysis: we specify a prior probability distribution (which can be subjective), and then update this distribution based on empirical data.

An example of when such approach would be necessary is a situation in which an experimenter bends a coin slightly and tosses it once, recording whether it comes up heads, and is then given the task of predicting the probability that the next flip comes up heads. After bending the coin, the true probability that the coin will come up heads is unknown; so the experimenter would need to make a decision (perhaps by looking at the shape of the coin) about what prior distribution to use. Incorporation of such subjective information might be important to get an accurate estimate of the probability.

Complexity

In general, model complexity involves a trade-off between simplicity and accuracy of the model. Occam's razor is a principle particularly relevant to modeling, its essential idea being that among models with roughly equal predictive power, the simplest one is the most desirable. While added complexity usually improves the realism of a model, it can make the model difficult to understand and analyze, and can also pose computational problems, including numerical instability. Thomas Kuhn argues that as science progresses, explanations tend to become more complex before a paradigm shift offers radical simplification.

For example, when modeling the flight of an aircraft, we could embed each mechanical part of the aircraft into our model and would thus acquire an almost white-box model of the system. However, the computational cost of adding such a huge amount of detail would effectively inhibit the usage of such a model. Additionally, the uncertainty would increase due to an overly complex system, because each separate part induces some amount of variance into the model. It is therefore usually appropriate to make some approximations to reduce the model to a sensible size. Engineers often can accept some approximations in order to get a more robust and simple model. For example, Newton's classical mechanics is an approximated model of the real world. Still, Newton's model is quite sufficient for most ordinary-life situations, that is, as long as particle speeds are well below the speed of light, and we study macro-particles only.

Training

Any model which is not pure white-box contains some parameters that can be used to fit the model to the system it is intended to describe. If the modeling is done by a neural network or other machine learning, the optimization of parameters is called *training*, while the optimization of model hyperparameters is called *tuning* and often uses cross-validation. In more conventional modeling through explicitly given mathematical functions, parameters are often determined by *curve fitting*.

Model Evaluation

A crucial part of the modeling process is the evaluation of whether or not a given mathematical model describes a system accurately. This question can be difficult to answer as it involves several different types of evaluation.

Fit to Empirical Data

Usually the easiest part of model evaluation is checking whether a model fits experimental measurements or other empirical data. In models with parameters, a common approach to test this fit is to split the data into two disjoint subsets: training data and verification data. The training data are used to estimate the model parameters. An accurate model will closely match the verification data even though these data were not used to set the model's parameters. This practice is referred to as cross-validation in statistics.

Defining a metric to measure distances between observed and predicted data is a useful tool of assessing model fit. In statistics, decision theory, and some economic models, a loss function plays a similar role.

While it is rather straightforward to test the appropriateness of parameters, it can be more difficult to test the validity of the general mathematical form of a model. In general, more mathematical tools have been developed to test the fit of statistical models than models involving differential equations. Tools from non-parametric statistics can sometimes be used to evaluate how well the data fit a known distribution or to come up with a general model that makes only minimal assumptions about the model's mathematical form.

Scope of the Model

Assessing the scope of a model, that is, determining what situations the model is applicable to, can be less straightforward. If the model was constructed based on a set of data, one must determine for which systems or situations the known data is a "typical" set of data.

The question of whether the model describes well the properties of the system between data points is called interpolation, and the same question for events or data points outside the observed data is called extrapolation.

As an example of the typical limitations of the scope of a model, in evaluating Newtonian classical mechanics, we can note that Newton made his measurements without advanced equipment, so he could not measure properties of particles travelling at speeds close to the speed of light. Likewise, he did not measure the movements of molecules and other small particles, but macro particles only. It is then not surprising that his model does not extrapolate well into these domains, even though his model is quite sufficient for ordinary life physics.

Philosophical Considerations

Many types of modeling implicitly involve claims about causality. This is usually (but not always) true of models involving differential equations. As the purpose of modeling is to increase our understanding of the world, the validity of a model rests not only on its fit to empirical observations, but also on its ability to extrapolate to situations or data beyond those originally described in the model. One can think of this as the differentiation between qualitative and quantitative predictions. One can also argue that a model is worthless unless it provides some insight which goes beyond what is already known from direct investigation of the phenomenon being studied.

An example of such criticism is the argument that the mathematical models of optimal foraging theory do not offer insight that goes beyond the common-sense conclusions of evolution and other basic principles of ecology.

Examples

- One of the popular examples in computer science is the mathematical models of various machines, an example is the deterministic finite automaton (DFA) which is defined as an abstract mathematical concept, but due to the deterministic nature of a DFA, it is implementable in hardware and software for solving various specific problems. For example, the following is a DFA M with a binary alphabet, which requires that the input contains an even number of 0s.

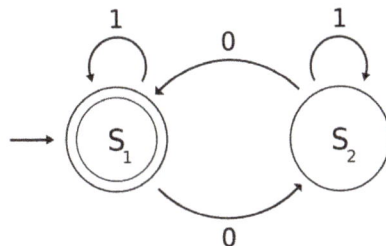

The state diagram for M

$M = (Q, \Sigma, \delta, q_0, F)$ where

- $Q = \{S_1, S_2\}$,

- $\Sigma = \{0, 1\}$,

- $q_0 = S_1$,

- $F = \{S_1\}$, and

- δ is defined by the following state transition table:

	0	1
S_1	S_2	S_1
S_2	S_1	S_2

The state S_1 represents that there has been an even number of 0s in the input so far, while S_2 signifies an odd number. A 1 in the input does not change the state of the automaton. When the input ends, the state will show whether the input contained an even number of 0s or not. If the input did contain an even number of 0s, M will finish in state S_1, an accepting state, so the input string will be accepted.

The language recognized by M is the regular language given by the regular expression $1^*(0 (1^*) 0 (1^*))^*$, where "*" is the Kleene star, e.g., 1^* denotes any non-negative number (possibly zero) of symbols "1".

- Many everyday activities carried out without a thought are uses of mathematical models. A geographical map projection of a region of the earth onto a small, plane surface is a model which can be used for many purposes such as planning travel.

- Another simple activity is predicting the position of a vehicle from its initial position, direction and speed of travel, using the equation that distance traveled is the product of time and speed. This is known as dead reckoning when used more formally. Mathematical modeling in this way does not necessarily require formal mathematics; animals have been shown to use dead reckoning.

- *Population Growth.* A simple (though approximate) model of population growth is the Malthusian growth model. A slightly more realistic and largely used population growth model is the logistic function, and its extensions.

- *Model of a particle in a potential-field.* In this model we consider a particle as being a point of mass which describes a trajectory in space which is modeled by a function giving its coordinates in space as a function of time. The potential field is given by a function $V: \mathbb{R}^3 \to \mathbb{R}$ and the trajectory, that is a function $r: \mathbb{R} \to \mathbb{R}^3$, is the solution of the differential equation:

$$-\frac{d^2\mathbf{r}(t)}{dt^2}m = \frac{\partial V[\mathbf{r}(t)]}{\partial x}\hat{\mathbf{x}} + \frac{\partial V[\mathbf{r}(t)]}{\partial y}\hat{\mathbf{y}} + \frac{\partial V[\mathbf{r}(t)]}{\partial z}\hat{\mathbf{z}},$$

that can be written also as:

$$m\frac{d^2\mathbf{r}(t)}{dt^2} = -\nabla V[\mathbf{r}(t)].$$

Note this model assumes the particle is a point mass, which is certainly known to be false in many cases in which we use this model; for example, as a model of planetary motion.

- *Model of rational behavior for a consumer*. In this model we assume a consumer faces a choice of n commodities labeled $1,2,...,n$ each with a market price p_1, $p_2,..., p_n$. The consumer is assumed to have an ordinal utility function U (ordinal in the sense that only the sign of the differences between two utilities, and not the level of each utility, is meaningful), depending on the amounts of commodities $x_1, x_2,..., x_n$ consumed. The model further assumes that the consumer has a budget M which is used to purchase a vector $x_1, x_2,..., x_n$ in such a way as to maximize $U(x_1, x_2,..., x_n)$. The problem of rational behavior in this model then becomes an optimization problem, that is:

$$\max U(x_1, x_2, \ldots, x_n)$$

subject to:

$$\sum_{i=1}^{n} p_i x_i \leq M.$$

$$x_i \geq 0 \quad \forall i \in \{1, 2, \ldots, n\}$$

This model has been used in a wide variety of economic contexts, such as in general equilibrium theory to show existence and Pareto efficiency of economic equilibria.

- *Neighbour-sensing model* explains the mushroom formation from the initially chaotic fungal network.

- *Computer science*: models in Computer Networks, data models, surface model,...

- *Mechanics*: movement of rocket model,...

Evaporation is widely employed in chemical, petrochemical, food, refrigeration, power plant and other allied industries to concentrate dilute aqueous solution to desired concentration so as to make end product suitable for further processing and marketable. It is an energy intensive operation as multitudes of thermal energy in the form of steam are used in it. In recent years, a radical increase in energy cost in relation to the capital equipment cost has caused a dramatic increase in the operating expense of the evaporators. This trend is quite likely to continue in future too, due to ever increasing rapid depletion in the reserve of fossil fuels and also their continuous consumption at alarming rates. Thus energy plays a dominating role in the design of evaporators. This emphasizes the need of energy in evaporators. one of the basic factor contributing to it is steam economy as it represents a measure of the quantity of water evaporated per unit of steam consumption. As a matter of fact, high values of steam economy are desirable in multiple effect evaporators so that the steam consumption is kept at the lowest level for a given evaporation of water.

Steam economy of a multiple effect evaporator having a specified feed arrangement depends on the number of effects, steam pressure, temperature, flow rate and concentration of feed; pressure in individual effects and physic-thermal properties of the solution. Obviously a change in any of this variable can affect the steam economy of the evaporator. Therefore, values of these variables are to be determined so that a evaporator with a given feed arrangement can operate for increased steam economy. This calls for a detailed investigation to study the parametric effect of above variables on steam economy of an evaporator for a given feed arrangement. This will also help plant engineers to revamp their existing evaporator by adjusting values of operating variables to attain improved steam economy.

It is to important to mention that in some of the cases a plant engineer doesn't have match flexibility to change the value of some of the above variable due to process constraints. For example concentration of the end product in a sugar solution, multiple effect evaporators is kept at the level of $60 \pm 5°$ Bx, otherwise seeding of sugar crystal will take place in the evaporator itself. Similarly, in the evaporation of caustic soda solution, concentration of the end product limited to a maximum of 50%. as a beyond this concentration the freezing point of the caustic soda solution start rising steeply. There is another situation with the captive caustic soda plant where relatively higher sodium contents in a caustic soda solution can be tolerated and the end product concentration is restricted only to 30-35%. Thus end product concentration of a solution may be fixed by the process technology, economics and other factors. For such system values of operating variable have to determine which can yield the possible steam economy of the evaporator. This necessitates the knowledge of the parametric effect of paramagnetic variable, namely; feed temperature, feed concentration, feed rate, pressure in the last effect and the steam pressure on the concentration of end product of an evaporator with the various feed arrangement.

Mathematical Modelling of Single Effect Evaporator

The equation describing the single effect evaporator is developed in the following manner. Component material balance on the solute and solvent are-

$$FX_f = LX_p \tag{1}$$
$$F(1 - X_f) = V + L(1 - X_p) \tag{2}$$

Respectively,

Where

F = feed rate kg/sec

L = concentrated liquid rate kg/sec

V = vapour rate kg/sec

X_f = mass fraction of solute in feed

X_p = mass fraction of solute in the product

Total material balance is given by-:

$$F = V + L \tag{3}$$

An enthalpy balance on the process stream yields-:

$$F.h_f + Q = V.H_v + L.h_p$$

$$or \qquad F.h_f + Q = (F - L)H_v + Lh_p + F.h_p - Fh_p$$
$$or \qquad F(h_f - h_p) + Q = (F - L)H_v - (F - L)h_p$$

$$or \quad F(h_f - h_p) + Q - (F - L)(H_v - H_p) = 0 \tag{4}$$

Where,

h_f = enthalpy of feed

h_p = enthalpy of product

H_v = enthalpy of vapour at the boiling point temperature of the feed, kJ/kg.

Q = rate of heat transfer across the tubes (from the steam to the thick liquor) kJ/sec
The enthalpy balance on the stream is given by-:

$$Q = V_0(H_o - h_o) = V_o \lambda_o \tag{5}$$

The rate of heat transfer Q is commonly approximated by use of the relationship-:

$$Q = UA(T_0 - T) \tag{6}$$

Where-:

U = overall heat transfer coefficient

A =surface area of the tubes available for heat transfer.

To = saturation temperature of the steam entering the chest.

T = boiling point temperature of thick liquor at the pressure of the vapour space

Evaporator

An evaporator is a device used to turn the liquid form of a chemical into its gaseous form. The liquid is evaporated, or vaporized, into a gas.

Uses

An evaporator is used in an air-conditioning system to allow a compressed cooling chemical, such as R-22 (Freon) or R-410A, to evaporate from liquid to gas while absorbing heat in the process. It can also be used to remove water or other liquids from mixtures. The process of evaporation is widely used to concentrate foods and chemicals as well as salvage solvents. In the concentration process, the goal of evaporation is to vaporize most of the water from a solution which contains the desired product. In the case of desalination of sea water or in Zero Liquid Discharge plants, the reverse purpose applies; evaporation removes the desirable drinking water from the undesired product, salt.

One of the most important applications of evaporation is in the food and beverage industry. Foods or beverages that need to last for a considerable amount of time or need to have certain consistency, like coffee, go through an evaporation step during processing.

In the pharmaceutical industry, the evaporation process is used to eliminate excess moisture, providing an easily handled product and improving product stability. Preservation of long-term activity or stabilization of enzymes in laboratories are greatly assisted by the evaporation process.

Another example of evaporation is in the recovery of sodium hydroxide in kraft pulping. Cutting down waste-handling cost is another major reason for large companies to use evaporation applications. Legally, all producers of waste must dispose of waste using methods compatible with environmental guidelines; these methods are costly. By removing moisture through vaporization, industry can greatly reduce the amount of waste product that must be processed.

Energetics

Water can be removed from solutions in ways other than evaporation, including membrane processes, liquid-liquid extractions, crystallization, and precipitation. Evaporation can be distinguished from some other drying methods in that the final product of evaporation is a concentrated liquid, not a solid. It is also relatively simple to use and understand since it has been widely used on a large scale, and many techniques are generally well known. In order to concentrate a product by water removal, an auxiliary phase is used which allows for easy transport of the solvent (water) rather than the solute. Water vapor is used as the auxiliary phase when concentrating non-volatile components, such as proteins and sugars. Heat is added to the solution, and part of the solvent is converted into vapor. Heat is the main tool in evaporation, and the process occurs more readily at high temperature and low pressures.

Heat is needed to provide enough energy for the molecules of the solvent to leave the solution and move into the air surrounding the solution. The energy needed can be expressed as an excess thermodynamic potential of the water in the solution. Leading to one of the biggest problems in industrial evaporation, the process requires enough energy to remove the water from the solution and to supply the heat of evaporation. When removing the water, more than 99% of the energy needed goes towards supplying the heat of evaporation. The need to overcome the surface tension of the solution also requires energy. The energy requirement of this process is very high because a phase transition must be caused; the water must go from a liquid to a vapor.

When designing evaporators, engineers must quantify the amount of steam needed for every mass unit of water removed when a concentration is given. An energy balance must be used based on an assumption that a negligible amount of heat is lost to the system's surroundings. The heat that needs to be supplied by the condensing steam will approximately equal the heat needed to vaporize the water. Another consideration is the size of the heat exchanger which affects the heat transfer rate.

Some common terms: A = heat transfer area, and q = overall heat transfer rate.

How an Evaporator Works

The solution containing the desired product is fed into the evaporator and passes across a heat source. The applied heat converts the water in the solution into vapor. The vapor is removed from the rest of the solution and is condensed while the non-concentrated solution is either fed into a second evaporator or is removed. The evaporator, as a machine, generally consists of four sections. The heating section contains the heating medium, which can vary. Steam is fed into this section. The most common medium consists of parallel tubes but others have plates or coils typically made from copper or aluminium. The concentrating and separating section removes the vapor being produced from the solution. The condenser condenses the separated vapor, then the vacuum or pump provides pressure to increase circulation.

Types of Evaporators used Today

Evaporator with SBT to eliminate bumping.

Natural/forced Circulation Evaporator

Natural circulation evaporators are based on the natural circulation of the product caused by the density differences that arise from heating. In an evaporator using tubing, after the water begins to boil, bubbles will rise and cause circulation, facilitating the separation of the liquid and the vapor at the top of the heating tubes. The amount of evaporation that takes place depends on the temperature difference between the steam and the solution.

Problems can arise if the tubes are not well-immersed in the solution. If this occurs, the system will be dried out and circulation compromised. In order to avoid this, forced circulation can be used by inserting a pump to increase pressure and circulation. Forced circulation occurs when hydrostatic head prevents boiling at the heating surface. A pump can also be used to avoid fouling that is caused by the boiling of liquid on the tubes; the pump suppresses bubble formation. Other problems are that the residing time is undefined and the consumption of steam is very high, but at high temperatures, good circulation is easily achieved.

Falling Film Evaporator

This type of evaporator is generally made of 4–8 m (13–26 ft) tubes enclosed by steam jackets. The uniform distribution of the solution is important when using this type of evaporator. The solution enters and gains velocity as it flows downward. This gain in velocity is attributed to the vapor being evolved against the heating medium, which flows downward as well. This evaporator is usually applied to highly viscous solutions, so it is frequently used in the chemical,sugar, food, and fermentation industries.

Rising Film (Long Tube Vertical) Evaporator

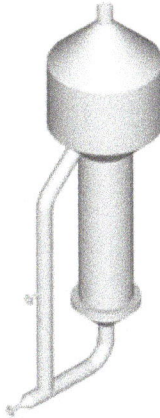

A rising film evaporator

In this type of evaporator, boiling takes place inside the tubes, due to heating made (usually by steam) outside the same. Submergence is therefore not desired; the creation of water vapor bubbles inside the tube creates an ascensional flow enhancing the heat transfer coefficient. This type of evaporator is therefore quite efficient, the disadvantage being to be prone to quick scaling of the internal surface of the tubes. This design is then usually applied to clear, non-salting solutions. Tubes are usually quite long, typically 4+ meters (13+ ft). Sometimes a small recycle is provided. Sizing this type of evaporator is usually a delicate task, since it requires a precise evaluation of the actual level of the process liquor inside the tubes. Recent applications tend to favor the falling-film pattern rather than rising-film.

Climbing and Falling-film Plate Evaporator

Climbing and falling-film plate evaporators have a relatively large surface area. The plates are usually corrugated and are supported by frame. During evaporation, steam flows through the channels formed by the free spaces between the plates. The steam alternately climbs and falls parallel to the concentrated liquid. The steam follows a co-current, counter-current path in relation to the liquid. The concentrate and the vapor are both fed into the separation stage where the vapor is sent to a condenser. This type of plate evaporator is frequently applied in the dairy and fermentation industries since they have spatial flexibility. A negative point of this type of evaporator is that it is limited in its ability to treat viscous or solid-containing products. There are other types of plate evaporators, which work with only climbing film.

Multiple-effect Evaporators

Unlike single-stage evaporators, these evaporators can be composed of up to seven evaporator stages (effects). The energy consumption for single-effect evaporators is very high and is most of the cost for an evaporation system. Putting together evaporators saves heat and thus requires less energy. Adding one evaporator to the original decreases energy consumption to 50%. Adding another effect reduces it to 33% and so on. A heat-saving-percent equation can be used to estimate how much one will save by adding a certain amount of effects.

The number of effects in a multiple-effect evaporator is usually restricted to seven because after that, the equipment cost approaches the cost savings of the energy-requirement drop.

There are two types of feeding that can be used when dealing with multiple-effect evaporators. Forward feeding takes place when the product enters the system through the first effect, which is at the highest temperature. The product is then partially concentrated as some of the water is transformed into vapor and carried away. It is then fed into the second effect which is slightly lower in temperature. The second effect uses the heated vapor created in the first stage as its heat source (hence the saving in energy expenditure). The combination of lower temperatures and higher viscosities in subsequent effects provides good conditions for treating heat-sensitive products, such as enzymes and proteins. In this system, an increase in the heating surface area of subsequent effects is required.

Another method is using backward feeding. In this process, the dilute products are fed into the last effect which has the lowest temperature and are transferred from effect to effect, with the temperature increasing. The final concentrate is collected in the hottest effect, which provides an advantage in that the product is highly viscous in the last stages, and so the heat transfer is better. Since some years there are also in operation multiple-effect vacuum evaporators with heat pump, well known to be energetically and technically more effective than systems with mechanical vapor recompression (MVR) because due to the lower boiling temperature they can handle highly corrosive liquids or which may form incrustations.

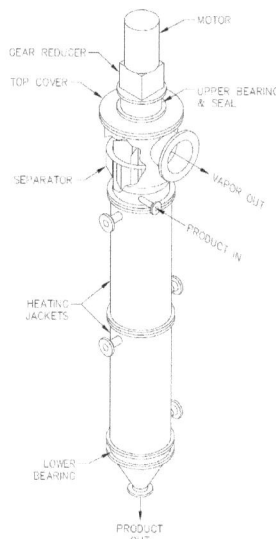

Agitated Thin / Wiped Film Evaporator Diagram

Agitated Thin Film Evaporators

Agitated thin-film evaporation has been very successful with difficult-to-handle products. Simply stated, the method quickly separates the volatile from the less volatile components using indirect heat transfer and mechanical agitation of the flowing product film under controlled conditions. The separation is normally made under vacuum conditions to maximize ΔT while maintaining the most favorable product temperature, and to maximize volatile stripping and recovery.

Problems

Technical problems can arise during evaporation, especially when the process is applied to the food industry. Some evaporators are sensitive to differences in viscosity and consistency of the dilute solution. These evaporators could work inefficiently because of a loss of circulation. The pump of an evaporator may need to be changed if the evaporator needs to be used to concentrate a highly viscous solution.

Fouling also occurs when hard deposits form on the surfaces of the heating mediums in the evaporators. In foods, proteins and polysaccharides can create such deposits that reduce the efficiency of heat transfer. Foaming can also create a problem since dealing with the excess foam can be costly in time and efficiency. Antifoam agents are to be used, but only a few can be used when food is being processed.

Corrosion can also occur when acidic solutions such as citrus juices are concentrated. The surface damage caused can shorten the long-life of evaporators. Quality and flavor of food can also suffer during evaporation. Overall, when choosing an evaporator, the qualities of the product solution need to be taken into careful consideration.

Marine Use

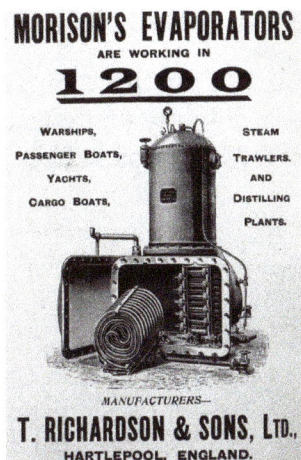

MORISON'S EVAPORATORS
ARE WORKING IN
1200
WARSHIPS, PASSENGER BOATS, YACHTS, CARGO BOATS, STEAM TRAWLERS, AND DISTILLING PLANTS.

MANUFACTURERS—
T. RICHARDSON & SONS, LTD.,
HARTLEPOOL, ENGLAND.

Large ships usually carry evaporating plants to produce fresh water, thus reducing their reliance on shore-based supplies. Steam ships must be able to produce high-quality distillate in order to maintain boiler-water levels. Diesel-engined ships often utilise waste heat as an energy source for producing fresh water. In this system, the engine-cooling water is passed through a heat exchanger, where it is cooled by concentrated seawater (brine). Because the cooling water (which is chemically treated fresh water) is at a temperature of 70–80 °C (158–176 °F), it would not be possible to flash off any water vapour unless the pressure in the heat exhanger vessel was dropped.

To alleviate this problem, a brine-air ejector venturi pump is used to create a vacuum inside the vessel. Partial evaporation is achieved, and the vapour passes through a demister before reaching the condenser section. Seawater is pumped through the condenser section to cool the vapour sufficiently to precipitate it. The distillate gathers in a tray, from where it is pumped to the storage tanks. A salinometer monitors salt content and diverts the flow of distillate from the storage tanks if the salt content exceeds the alarm limit. Sterilisation is carried out after the evaporator.

Evaporators are usually of the shell-and-tube type (known as an Atlas Plant) or of the plate type (such as the type designed by Alfa Laval). Temperature, production and vacuum are controlled by regulating the system valves. Seawater temperature can interfere with production, as can fluctuations in engine load. For this reason, the evaporator is adjusted as seawater temperature changes, and shut down altogether when the ship is manoeuvring. An alternative in some vessels, such as naval ships and passenger ships, is the use of the reverse osmosis principle for fresh-water production, instead of using evaporators.

The chief factor influencing the economy of an evaporator system is the number of effects. By increasing the number of effects we can increase the economy of an evaporator system. The first effect of a multiple effect evaporator is the effect to which the raw steam is fed, vapors obtained from first effect act as a heating medium for another effect.

Different Types of Feed Arrangement of Multiple Effect Evaporators -:

1) FORWARD FEED ARRANGEMENT : In this arrangement the feed and steam introduced in the first effect and pressure in the first effect is highest and pressure in last effect is minimum, so transfer of feed from one effect to another can be done without pump.

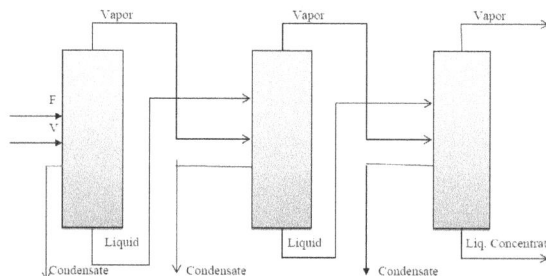

Forward feed arrangement in Multi effect Evaporator

2) BACKWARD FEED ARRANGEMENT: In this arrangement feed is introduced in last effect and steam is introduced in first effect. For transfer of feed, it requires pump since the flow is from low pressure to higher pressure. Concentrated liquid is obtained in first effect.

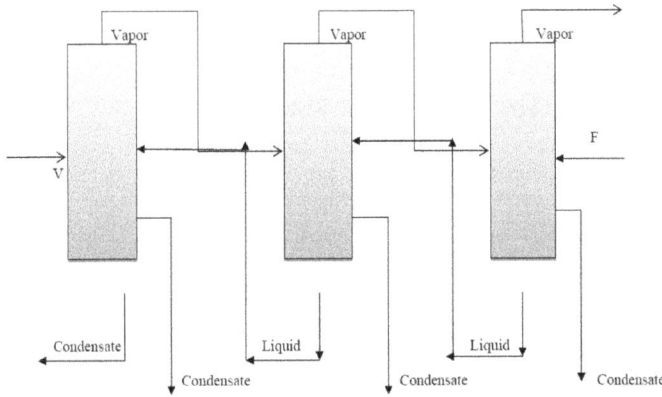

Backward feed arrangement in Multi effect Evaporator

3) MIXED FEED ARRANGEMENT: In this arrangement feed is introduced in intermediate effect, flows in forward feed to the end of the series and is then pumped back to the first effect for final concentration. This permits the final evaporation to be done at the highest temperature.

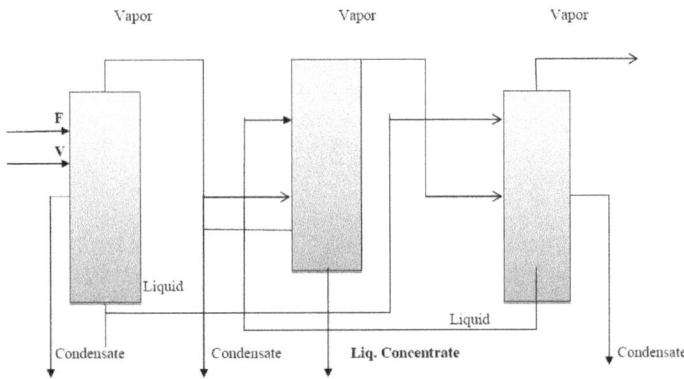

Mixed feed arrangement in Multi effect Evaporator

Analysis of Multiple Effect Evaporator System:

With the increasing trend in the cost of coal, fuel oil it becomes to use the vapors of previous effect in the steam chest of following effect. This requires the multiple effect evaporator system.As the number of effect increases the steam economy increases on the other side capital cost will be more.

There is economic balance between the fixed cost and the operating cost so that one can select the optimum number of effects.

Fixed Cost:

According to Coston and Lindey, the annual fixed cost of a multiple effect evaporator is approximately proportional to the 0.75 power of the number of effects.

$$V_1 = C_1 / A * N^{0.75}$$

The estimated cost of a single effect (C_1) can be obtained from a number of sources once the heat transfer surfaces requirements are known.

Operating Cost:

Operating costs can be divided in to steam cost and all operating costs (labour, cooling water, power and maintenance) such that

$$V_1 = h * W * C_2 / S + V_0$$

H = operating time (hr/yr)

W = evaporation rate (Kg/hr)

C_2 = cost of steam (Rs/Kg)

S = steam economy

V_0 = all operating cost other than the cost of steam (Rs/Yr)

Steam economy can be expressed as:

$$S = S_1 + S_1 S_2 + S_1 S_2^2 + \ldots\ldots\ldots + S_1 S_2^{N-1}$$

$$S = S_1 (1 - S_2^N) / (1 - S_2)$$

Thus V_2 becomes

$$V_2 = (1 - S_2) h * W * C_2 / S_1 (1 - S_2^N) + V_0$$

Total cost

$$V_T = V_1 + V_2$$

$$V_T = \; = C_1 / A * N^{0.75} + (1 - S_2) h * W * C_2 / S_1 (1 - S_2 N) + V_0$$

For minimum cost $\Delta V_T / \Delta N = 0$

$$0 = C_1 / N\{(N+1)^{0.75} - (N^{0.75})\} + (1 - S_2) h * W * C_2 / S_1 \{[1 / (1 - S_2^{N+1})] - [1 / (1 - S_2^N)]\}$$

$$A h w c_2 / C_1 = \{[(N+1)^{0.75} - (N^{0.75})] * S_1 (1 - S_2^N)(1 - S_2^{N+1})\} / (1 - S_2)^2 * S_2^N$$

Modeling of Multiple Effect Evaporator System

Multiple effect evaporators involve a large number of state and design –variables. A change in any variable can upset the operation of the evaporator. To achieve the goal of energy conservation in multiple effect evaporators, it is necessary to know how does steam economy alter with changes in operating variables for a given end product concentration. To quantise the changes in steam economy, a functional relationship correlating it with variables should be developed. For this, it is necessary to identify all the variables which affect the steam economy of a multiple effect evaporator.

Variables of a Multiple Effect Evaporator:

In a evaporator, the variables can be classified as geometrical-operating, and self balancing variables. As regards the geometrical variable, it is the area of heat transfer surface in each effect of an evaporator. Hence, N-effect evaporator will have N number of geometrical variables.

From industrial practices.we know that there are some operating variables which plant engineer can change them independently to annual any imbalance in the operation of an evaporator. They include: feed temperature, feed concentration,feed flow rate,and steam temperature (pressure), saturation temperature (pressure) in the last effect. Feed arrangement (forward/backward/mixed) is also one of the operating variables. Thus, total number of operating variables is six.

As regards the vapour and liquid streams from each effect of a multiple effect evaporator, they cannot be changed independently by a plant engineer. Therefore, they are self balancing streams. The variables associated with these streams are: flow rate, temperature and concentration of liquid streams ; and saturation temperature(pressure) of each effect. However, temperature of vapour stream equals to the temperature of liquid stream. In this way, for N-effect effect evaporator the number of self balancing variables becomes 5N. It is important to point out here that the saturation temperature(pressure) of the last effect, has already been taken in to account as an operating variable. Therefore, it cannot be considered as a self- balancing variable. Flow rate of steam to the first effect is the another self-balancing variable whose value is usually not altered. Thus the total number of net self balancing variables for N- effect evaporator, becomes 5N [=5N-1+1].

The summation of geometrical, operating and self -balancing variables gives the total number of variables in an evaporator. They are equal to 6N+6[=N+6+5N].

Mathematical Model:

A mathematical model of a multiple effect evaporator is a relationship amongst the geometrical, operating and self -balancing variables. This can be obtained from the equations of material balance, energy balance, heat transfer rate, and boiling point rise.

For the simplicity of the mathematical model, following assumptions have been made in this analysis:

1. The vapours entering in to steam chest of respective effects are at their saturation temperature.

2. There is no sub cooling of the condensate from different steam chests.

3. Condensation of vapour in steam chest occurs at constant pressure.

4. There is no carry- over of liquid droplets with vapors leaving the respective effects.

5. There is no heat dissipation to surroundings.

6. Heat transfer surface does not undergo fouling.

Design of Multiple Effect Evaporator Without Boiling Point Elevation for Forward Feed:

Equations are developed for the case where boiling point elevations are negligible,also the effect of composition on liquid enthalpy is neglected. The equations so obtained are generalized for the case where boiling point elevations cannot be neglected. For definiteness, forward feeds are employed.

Specifications: F, X_F,T_F,T_0,P_0,P_3(ort$_3$),X_3(orl$_3$),U_1,U_2,U_3,equal areas, forward feed, negligible boiling point elevations.

To find : V_0, T_1,L_1,T_2,L_2 and A

Actually,four additional dependent variables exist; namely: V_1,V_2,X_1,X_2

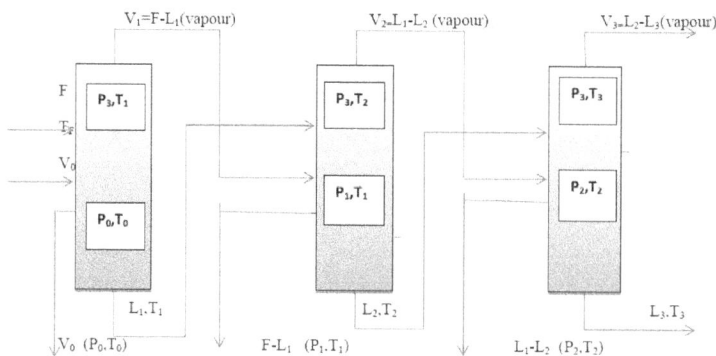

Triple Effect Evaporator with Forward feed Arrangement

Equations of energy balance and heat transfer rate for all three effects are-:

Effect -1:

Enthalpy balance:

$$F(h_f - h_1) + V_0\lambda_0 = (F - L_1)\lambda_1$$

Rate equation:

$$U_1 A(T_0 - T_1) = V_0\lambda$$

Effect -2:

Enthalpy balance:

$$L_1(h_1 - h_2) + (F - L_1)\lambda = (L_1 - L_2)\lambda_2$$

Rate equation:

$$U_2 A(T_1 - T_2) = (F - L_1)\lambda_1$$

Effect -3 :

Enthalpy balance:

$$L_2(h_2 - h_3) + (L_1 - L_2)\lambda_2 = (L_2 - L_3)\lambda_3$$

Rate equation:

$$U_3 A(T_2 - T_3) = (L_1 - L_2)\lambda_2$$

The above equation describing the triple effect evaporator system constitute a set of nonlinear algebraic equations that may be solved in variety of way, one of which is Newton raphson method.

Design of Multiple Effect Evaporator with Boiling Point Rise for Forward Feed:

Most solutions that are concentrated in evaporators are mixtures of water and non-volatile salt. The boiling temperature of the solution some times depend on the salt concentration. The difference between the temperature(T) of boiling solution and the temperature of boiling water (pure) at the same pressure is known as boiling point rise(BPR).thus

$$T = T_W + BPR$$

BPR is a function of the solute concentration. A graph called the Dhuring chart is commomnly used to determine BPR.

Calculation procedure: Specifications:

F, X_F, T_F, thick liquid composition(X_p), P, saturated steam pressure (P_o), heat capacity(C_p), overall heat transfer coefficient (U).

To find: Heat transfer area(A)

Step-1:

Corresponding to evaporator temperature find out the boiling point of pure water, T_w.

Step-2:

From the figures/empirical correlation determine the solution temperature/BPR at T_w, X_p. This temperature is also the temperature of the superheated water vapour leaving the evaporator.

Step-3:

A total mass balance and a component material balance are used to calculate the flow rates.

$$F = V + L$$

$$X_F F = L X_p$$

Step-4:

Calculate /determine the enthalpies of three process streams

1) Enthalpy of water vapour from the superheated steam tables by solution temperature and pressure of the evaporator.

2) The enthalpy of the solution can be calculated from there heat capacities. H = $C_p(T\text{-}T_R)$

It can also be calculated by the plots or empirical correlations.

3) Latent heat of vaporization taken from steam table at corresponding steam pressure.

Step-5:

Now, write the enthalpy balance

$$F(h_f - h_p) + V_0 \lambda_0 = (F - L)(H_V - h_p)$$

$$Or, \ V_0 = [(F - L)(H_V - h_p) - F(h_f - h_p)]/ \lambda_0$$

Step-6:

$V_0 \lambda_0 = UA(T_0 - T_1)$.

$A = V_0\lambda_0/U(T_0\text{-}T_1)$

Equations for all three effects are-:

Effect -1:

$F[h(T_f,X_f)-h(T_1,X_1)]+ V_0\ \lambda_0\ -(F-L_1)[H(T_1)-h(T_1,X_1)]=0$

$U_1A_1\ (T_0\ -T_1) - V_0\ \lambda_0 =0$

$X_F\ F - L_1\ X_1 = 0$

Effect -2:

$L_1[h(T_1,X_1)-h(T_2)]+ (F-L_1)[H(T_1)-h(T_1)]- (L_1-L_2)[H(T_2)-h(T_2)]=0$

$U_2\ A_2\ (T_1\ -T_2)- (F-L_1)[H(T_1)-h(T_1)] = 0$

$L_1\ X_1\ - L_2\ X_2 = 0$

Effect -3:

$L_2[h(T_2,X2)-h(T_3,\ X_3)] + (L_1-L_2)[H(T_2)-h(T_2)]- (L_2-L_3)[H(T_3)-h\ (T_3,\ X_3)]= 0$

$U_3\ A_3\ (T_2\ -T_3)-(L_1-L_2)[H-h]=0$

$L_2\ X_2\ - L_3\ X_3 = 0$

Newton's Method

In numerical analysis, Newton's method (also known as the Newton–Raphson method), named after Isaac Newton and Joseph Raphson, is a method for finding successively better approximations to the roots (or zeroes) of a real-valued function.

$$x : f(x) = 0.$$

The Newton–Raphson method in one variable is implemented as follows:

The method starts with a function f defined over the real numbers x, the function's derivative f', and an initial guess x_0 for a root of the function f. If the function satisfies the assumptions made in the derivation of the formula and the initial guess is close, then a better approximation x_1 is

$$x_1 = x_0 - \frac{f(x_0)}{f'(x_0)}.$$

Geometrically, $(x_1, 0)$ is the intersection of the x-axis and the tangent of the graph of f at $(x_0, f(x_0))$.

The process is repeated as

$$x_{n+1} = x_n - \frac{f(x_n)}{f'(x_n)}$$

until a sufficiently accurate value is reached.

This algorithm is first in the class of Householder's methods, succeeded by Halley's method. The method can also be extended to complex functions and to systems of equations.

Description

The idea of the method is as follows: one starts with an initial guess which is reasonably close to the true root, then the function is approximated by its tangent line (which can be computed using the tools of calculus), and one computes the x-intercept of this tangent line (which is easily done with elementary algebra). This x-intercept will typically be a better approximation to the function's root than the original guess, and the method can be iterated.

Suppose $f: [a, b] \to \mathbb{R}$ is a differentiable function defined on the interval $[a, b]$ with values in the real numbers $\mathbb{R}$. The formula for converging on the root can be easily derived. Suppose we have some current approximation x_n. Then we can derive the formula for a better approximation, x_{n+1} by referring to the diagram on the right. The equation of the tangent line to the curve $y = f(x)$ at the point $x = x_n$ is

$$y = f'(x_n)(x - x_n) + f(x_n),$$

where f' denotes the derivative of the function f.

The x-intercept of this line (the value of x such that $y = 0$) is then used as the next approximation to the root, x_{n+1}. In other words, setting y to zero and x to x_{n+1} gives

$$0 = f'(x_n)(x_{n+1} - x_n) + f(x_n).$$

Solving for x_{n+1} gives

$$x_{n+1} = x_n - \frac{f(x_n)}{f'(x_n)}.$$

We start the process off with some arbitrary initial value x_0. (The closer to the zero, the better. But, in the absence of any intuition about where the zero might lie, a "guess and check" method might narrow the possibilities to a reasonably small interval by appealing to the intermediate value theorem.) The method will usually converge, provided this initial guess is close enough to the unknown zero, and that $f'(x_0) \neq 0$. Furthermore, for a zero of multiplicity 1, the convergence is at least quadratic in a neighbourhood of the zero, which intuitively means that the number of correct digits roughly at least doubles in every step.

The Householder's methods are similar but have higher order for even faster convergence. However, the extra computations required for each step can slow down the overall performance relative to Newton's method, particularly if f or its derivatives are computationally expensive to evaluate.

History

The name "Newton's method" is derived from Isaac Newton's description of a special case of the method in *De analysi per aequationes numero terminorum infinitas* (written in 1669, published in 1711 by William Jones) and in *De metodis fluxionum et serierum infinitarum* (written in 1671, translated and published as *Method of Fluxions* in 1736 by John Colson). However, his method differs substantially from the modern method given above: Newton applies the method only to polynomials. He does not compute the successive approximations x_n, but computes a sequence of polynomials, and only at the end arrives at an approximation for the root x. Finally, Newton views the method as purely algebraic and makes no mention of the connection with calculus. Newton may have derived his method from a similar but less precise method by Vieta. The essence of Vieta's method can be found in the work of the Persian mathematician Sharaf al-Din al-Tusi, while his successor Jamshīd al-Kāshī used a form of Newton's method to solve $x^p - N = 0$ to find roots of N (Ypma 1995). A special case of Newton's method for calculating square roots was known much earlier and is often called the Babylonian method.

Newton's method was used by 17th-century Japanese mathematician Seki Kōwa to solve single-variable equations, though the connection with calculus was missing.

Newton's method was first published in 1685 in *A Treatise of Algebra both Historical and Practical* by John Wallis. In 1690, Joseph Raphson published a simplified description in *Analysis aequationum universalis*. Raphson again viewed Newton's method purely as an algebraic method and restricted its use to polynomials, but he describes the method in terms of the successive approximations x_n instead of the more complicated sequence of polynomials used by Newton. Finally, in 1740, Thomas Simpson described Newton's method as an iterative method for solving general nonlinear equations using calculus, essentially giving the description above. In the same publication, Simpson also gives the generalization to systems of two equations and notes that Newton's method can be used for solving optimization problems by setting the gradient to zero.

Arthur Cayley in 1879 in *The Newton-Fourier imaginary problem* was the first to notice the difficulties in generalizing Newton's method to complex roots of polynomials with degree greater than 2 and complex initial values. This opened the way to the study of the theory of iterations of rational functions.

Practical Considerations

Newton's method is an extremely powerful technique—in general the convergence is quadratic: as the method converges on the root, the difference between the root and the approximation is squared (the number of accurate digits roughly doubles) at each step. However, there are some difficulties with the method.

Difficulty in Calculating Derivative of a Function

Newton's method requires that the derivative be calculated directly. An analytical expression for the derivative may not be easily obtainable and could be expensive to evaluate. In these situations, it may be appropriate to approximate the derivative by using the slope of a line through two nearby points on the function. Using this approximation would result in something like the secant method whose convergence is slower than that of Newton's method.

Failure of the Method to Converge to the Root

It is important to review the proof of quadratic convergence of Newton's Method before implementing it. Specifically, one should review the assumptions made in the proof. For situations where the method fails to converge, it is because the assumptions made in this proof are not met.

Overshoot

If the first derivative is not well behaved in the neighborhood of a particular root, the method may overshoot, and diverge from that root. An example of a function with one root, for which the derivative is not well behaved in the neighborhood of the root, is

$$f(x) = |x|^a, \quad 0 < a < \tfrac{1}{2}$$

for which the root will be overshot and the sequence of x will diverge. For $a = 1/2$, the root will still be overshot, but the sequence will oscillate between two values. For $1/2 < a < 1$, the root will still be overshot but the sequence will converge, and for $a \geq 1$ the root will not be overshot at all.

In some cases, Newton's method can be stabilized by using successive over-relaxation, or the speed of convergence can be increased by using the same method.

Stationary Point

If a stationary point of the function is encountered, the derivative is zero and the method will terminate due to division by zero.

Poor Initial Estimate

A large error in the initial estimate can contribute to non-convergence of the algorithm. To overcome this problem one can often linearise the function that is being optimized using calculus, logs, differentials, or even using evolutionary algorithms, such as the Stochastic Funnel Algorithm. Good initial estimates lie close to the final globally optimal parameter estimate. In nonlinear regression, the sum of squared errors (SSE) is only "close to" parabolic in the region of the final parameter estimates. Initial estimates found here will allow the Newton-Raphson method to quickly converge. It is only here that the Hessian matrix of the SSE is positive and the first derivative of the SSE is close to zero.

Mitigation of Non-convergence

In a robust implementation of Newton's method, it is common to place limits on the number of iterations, bound the solution to an interval known to contain the root, and combine the method with a more robust root finding method.

Slow Convergence for Roots of Multiplicity Greater than 1

If the root being sought has multiplicity greater than one, the convergence rate is merely linear (errors reduced by a constant factor at each step) unless special steps are taken. When there are two or more roots that are close together then it may take many iterations before the iterates get close enough to one of them for the quadratic convergence to be apparent. However, if the multiplicity m of the root is known, the following modified algorithm preserves the quadratic convergence rate:

$$x_{n+1} = x_n - m \frac{f(x_n)}{f'(x_n)}.$$

This is equivalent to using successive over-relaxation. On the other hand, if the multiplicity m of the root is not known, it is possible to estimate m after carrying out one or two iterations, and then use that value to increase the rate of convergence.

Analysis

Suppose that the function f has a zero at α, i.e., $f(\alpha) = 0$, and f is differentiable in a neighborhood of α.

If f is continuously differentiable and its derivative is nonzero at α, then there exists a neighborhood of α such that for all starting values x_0 in that neighborhood, the sequence $\{x_n\}$ will converge to α.

If the function is continuously differentiable and its derivative is not 0 at α and it has a second derivative at α then the convergence is quadratic or faster. If the second derivative is not 0 at α then the convergence is merely quadratic. If the third derivative exists and is bounded in a neighborhood of α, then:

$$\Delta x_{i+1} = \frac{f''(\alpha)}{2 f'(\alpha)} (\Delta x_i)^2 + O(\Delta x_i)^3,$$

where

$$\Delta x_i \triangleq x_i - \alpha.$$

If the derivative is 0 at α, then the convergence is usually only linear. Specifically, if f is twice continuously differentiable, $f'(\alpha) = 0$ and $f''(\alpha) \neq 0$, then there exists a neighborhood of α such that for all starting values x_0 in that neighborhood, the sequence of iterates converges linearly, with rate $\log_{10} 2$. Alternatively if $f'(\alpha) = 0$ and $f'(x) \neq 0$ for $x \neq \alpha$, x in a neighborhood U of α, α being a zero of multiplicity r, and if $f \in C^r(U)$ then there exists a neighborhood of α such that for all starting values x_0 in that neighborhood, the sequence of iterates converges linearly.

However, even linear convergence is not guaranteed in pathological situations.

In practice these results are local, and the neighborhood of convergence is not known in advance. But there are also some results on global convergence: for instance, given a right neighborhood U_+ of α, if f is twice differentiable in U_+ and if $f' \neq 0$, $f \cdot f'' > 0$ in U_+, then, for each x_0 in U_+ the sequence x_k is monotonically decreasing to α.

Proof of Quadratic Convergence for Newton's Iterative Method

According to Taylor's theorem, any function $f(x)$ which has a continuous second derivative can be represented by an expansion about a point that is close to a root of $f(x)$. Suppose this root is α. Then the expansion of $f(\alpha)$ about x_n is:

$$f(\alpha) = f(x_n) + f'(x_n)(\alpha - x_n) + R_1$$

where the Lagrange form of the Taylor series expansion remainder is

$$R_1 = \frac{1}{2!} f''(\xi_n)(\alpha - x_n)^2,$$

where ξ_n is in between x_n and α.

Since α is the root, (1) becomes:

$$0 = f(\alpha) = f(x_n) + f'(x_n)(\alpha - x_n) + \tfrac{1}{2} f''(\xi_n)(\alpha - x_n)^2$$

Dividing equation (2) by $f'(x_n)$ and rearranging gives

$$\frac{f(x_n)}{f'(x_n)} + (\alpha - x_n) = \frac{-f''(\xi_n)}{2f'(x_n)}(\alpha - x_n)^2$$

Remembering that x_{n+1} is defined by

$$x_{n+1} = x_n - \frac{f(x_n)}{f'(x_n)},$$

one finds that

$$\underbrace{\alpha - x_{n+1}}_{\varepsilon_{n+1}} = \frac{-f''(\xi_n)}{2f'(x_n)} \underbrace{(\alpha - x_n)^2}_{\varepsilon_n}.$$

That is,

$$\varepsilon_{n+1} = \frac{-f''(\xi_n)}{2f'(x_n)} \cdot \varepsilon_n^2.$$

Taking absolute value of both sides gives

$$\left| \varepsilon_{n+1} \right| = \frac{\left| f''(\xi_n) \right|}{2 \left| f'(x_n) \right|} \cdot \varepsilon_n^2.$$

Equation shows that the rate of convergence is quadratic if the following conditions are satisfied:

1. $f'(x) \neq 0$; for all $x \in I$, where I is the interval $[\alpha - r, \alpha + r]$ for some $r \geq |\alpha - x_0|$;

2. $f''(x)$ is continuous, for all $x \in I$;

3. x_0 sufficiently close to the root α.

The term *sufficiently* close in this context means the following:

a. Taylor approximation is accurate enough such that we can ignore higher order terms;

b. $\dfrac{1}{2} \left| \dfrac{f''(x_n)}{f'(x_n)} \right| < C \left| \dfrac{f''(\alpha)}{f'(\alpha)} \right|$, for some $C < \infty$;

c. $C \left| \dfrac{f''(\alpha)}{f'(\alpha)} \right| \varepsilon_n < 1$, for $n \in \mathbb{Z}$, $n \geq 0$ and C satisfying condition b.

Finally, **(6)** can be expressed in the following way:

$$\left| \varepsilon_{n+1} \right| \leq M \varepsilon_n^2$$

where M is the supremum of the variable coefficient of ε_n^2 on the interval I defined in condition 1, that is:

$$M = \sup_{x \in I} \frac{1}{2} \left| \frac{f''(x)}{f'(x)} \right|.$$

The initial point x_0 has to be chosen such that conditions 1 to 3 are satisfied, where the third condition requires that $M |\varepsilon_0| < 1$.

Basins of Attraction

The basins of attraction—the regions of the real number line such that within each region iteration from any point leads to one particular root—can be infinite in number and arbitrarily small. For example, for the function $f(x) = x^3 - 2x^2 - 11x + 12 = (x - 4)(x - 1)(x + 3)$, the following initial conditions are in successive basins of attraction:

2.35287527	converges to	4;
2.35284172	converges to	-3;
2.35283735	converges to	4;
2.352836327	converges to	-3;
2.352836323	converges to	1.

Failure Analysis

Newton's method is only guaranteed to converge if certain conditions are satisfied. If the assumptions made in the proof of quadratic convergence are met, the method will converge. For the following subsections, failure of the method to converge indicates that the assumptions made in the proof were not met.

Bad Starting Points

In some cases the conditions on the function that are necessary for convergence are satisfied, but the point chosen as the initial point is not in the interval where the method converges. This can happen, for example, if the function whose root is sought approaches zero asymptotically as x goes to ∞ or $-\infty$. In such cases a different method, such as bisection, should be used to obtain a better estimate for the zero to use as an initial point.

Iteration Point is Stationary

Consider the function:

$$f(x) = 1 - x^2.$$

It has a maximum at $x = 0$ and solutions of $f(x) = 0$ at $x = \pm 1$. If we start iterating from the stationary point $x_0 = 0$ (where the derivative is zero), x_1 will be undefined, since the tangent at $(0,1)$ is parallel to the x-axis:

$$x_1 = x_0 - \frac{f(x_0)}{f'(x_0)} = 0 - \frac{1}{0}.$$

The same issue occurs if, instead of the starting point, any iteration point is stationary. Even if the derivative is small but not zero, the next iteration will be a far worse approximation.

Starting Point Enters a Cycle

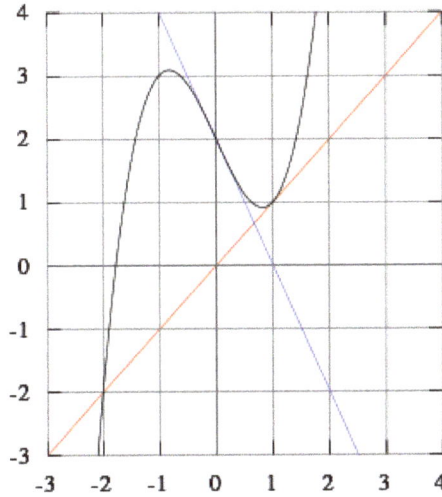

The tangent lines of $x^3 - 2x + 2$ at 0 and 1 intersect the x-axis at 1 and 0 respectively, illustrating why Newton's method oscillates between these values for some starting points.

For some functions, some starting points may enter an infinite cycle, preventing convergence. Let

$$f(x) = x^3 - 2x + 2$$

and take 0 as the starting point. The first iteration produces 1 and the second iteration returns to 0 so the sequence will alternate between the two without converging to a root. In fact, this 2-cycle is stable: there are neighborhoods around 0 and around 1 from which all points iterate asymptotically to the 2-cycle (and hence not to the root of the function). In general, the behavior of the sequence can be very complex. The real solution of this equation is −1.76929235....

Derivative Issues

If the function is not continuously differentiable in a neighborhood of the root then it is possible that Newton's method will always diverge and fail, unless the solution is guessed on the first try.

Derivative does not Exist at Root

A simple example of a function where Newton's method diverges is trying to find the cube root of zero. The cube root is continuous and infinitely differentiable, except for x = 0, where its derivative is undefined:

$$f(x) = \sqrt[3]{x}.$$

For any iteration point x_n, the next iteration point will be:

$$x_{n+1} = x_n - \frac{f(x_n)}{f'(x_n)} = x_n - \frac{x_n^{\frac{1}{3}}}{\frac{1}{3}x_n^{\frac{1}{3}-1}} = x_n - 3x_n = -2x_n.$$

The algorithm overshoots the solution and lands on the other side of the y-axis, farther away than it initially was; applying Newton's method actually doubles the distances from the solution at each iteration.

In fact, the iterations diverge to infinity for every $f(x) = |x|^\alpha$, where $0 < \alpha < 1/2$. In the limiting case of $\alpha = 1/2$ (square root), the iterations will alternate indefinitely between points x_0 and $-x_0$, so they do not converge in this case either.

Discontinuous Derivative

If the derivative is not continuous at the root, then convergence may fail to occur in any neighborhood of the root. Consider the function

$$f(x) = \begin{cases} 0 & \text{if } x = 0, \\ x + x^2 \sin\dfrac{2}{x} & \text{if } x \neq 0. \end{cases}$$

Its derivative is:

$$f'(x) = \begin{cases} 1 & \text{if } x = 0, \\ 1 + 2x\sin\dfrac{2}{x} - 2\cos\dfrac{2}{x} & \text{if } x \neq 0. \end{cases}$$

Within any neighborhood of the root, this derivative keeps changing sign as x approaches 0 from the right (or from the left) while $f(x) \geq x - x^2 > 0$ for $0 < x < 1$.

So $f(x)/f'(x)$ is unbounded near the root, and Newton's method will diverge almost everywhere in any neighborhood of it, even though:

- the function is differentiable (and thus continuous) everywhere;
- the derivative at the root is nonzero;
- f is infinitely differentiable except at the root; and
- the derivative is bounded in a neighborhood of the root (unlike $f(x)/f'(x)$).

Non-quadratic Convergence

In some cases the iterates converge but do not converge as quickly as promised. In these cases simpler methods converge just as quickly as Newton's method.

Zero Derivative

If the first derivative is zero at the root, then convergence will not be quadratic. Let

$$f(x) = x^2$$

then $f'(x) = 2x$ and consequently

$$x - \frac{f(x)}{f'(x)} = \frac{x}{2}.$$

So convergence is not quadratic, even though the function is infinitely differentiable everywhere.

Similar problems occur even when the root is only "nearly" double. For example, let

$$f(x) = x^2(x-1000) + 1.$$

Then the first few iterations starting at $x_0 = 1$ are

$x_0 = 1$

$x_1 = 0.500250376\ldots$

$x_2 = 0.251062828\ldots$

$x_3 = 0.127507934\ldots$

$x_4 = 0.067671976\ldots$

$x_5 = 0.041224176\ldots$

$x_6 = 0.032741218\ldots$

$x_7 = 0.031642362\ldots$

it takes six iterations to reach a point where the convergence appears to be quadratic.

No Second Derivative

If there is no second derivative at the root, then convergence may fail to be quadratic. Let

$$f(x) = x + x^{\frac{4}{3}}.$$

Then

$$f'(x) = 1 + \tfrac{4}{3} x^{\frac{1}{3}}.$$

and

$$f''(x) \quad -x^{-}$$

except when $x = 0$ where it is undefined. Given x_n,

$$x_{n+1} = x_n - \frac{f(x_n)}{f'(x_n)} = \frac{\frac{1}{3}x_n^{\frac{4}{3}}}{1 + \frac{4}{3}x_n^{\frac{1}{3}}}$$

which has approximately 4/3 times as many bits of precision as x_n has. This is less than the 2 times as many which would be required for quadratic convergence. So the convergence of Newton's method (in this case) is not quadratic, even though: the function is continuously differentiable everywhere; the derivative is not zero at the root; and f is infinitely differentiable except at the desired root.

Generalizations

Complex Functions

When dealing with complex functions, Newton's method can be directly applied to find their zeroes. Each zero has a basin of attraction in the complex plane, the set of all starting values that cause the method to converge to that particular zero. These sets can be mapped as in the image shown. For many complex functions, the boundaries of the basins of attraction are fractals.

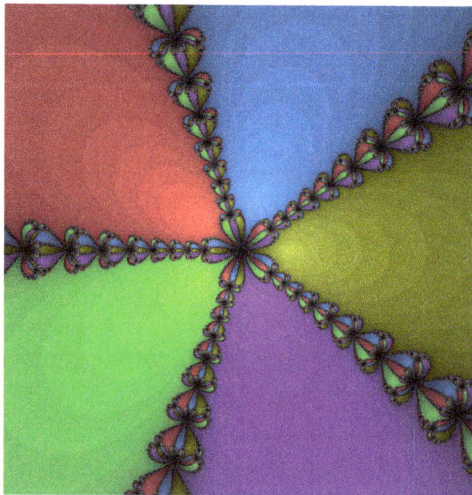

Basins of attraction for $x^5 - 1 = 0$; darker means more iterations to converge.

In some cases there are regions in the complex plane which are not in any of these basins of attraction, meaning the iterates do not converge. For example, if one uses a real initial condition to seek a root of $x^2 + 1$, all subsequent iterates will be real numbers and so the iterations cannot converge to either root, since both roots are non-real. In this case almost all real initial conditions lead to chaotic behavior, while some initial conditions iterate either to infinity or to repeating cycles of any finite length.

Curt McMullen has shown that for any possible purely iterative algorithm similar to Newton's Method, the algorithm will diverge on some open regions of the complex plane when applied to some polynomial of degree 4 or higher. However, McMullen gave a generally convergent algorithm for polynomials of degree 3.

Nonlinear Systems of Equations

k Variables, k Functions

One may also use Newton's method to solve systems of k (nonlinear) equations, which amounts to finding the zeroes of continuously differentiable functions $F : \mathbb{R}^k \to \mathbb{R}^k$. In the formulation given above, one then has to left multiply with the inverse of the $k \times k$ Jacobian matrix $J_F(x_n)$ instead of dividing by $f'(x_n)$.

Rather than actually computing the inverse of this matrix, one can save time by solving the system of linear equations

$$J_F(x_n)(x_{n+1} - x_n) = -F(x_n)$$

for the unknown $x_{n+1} - x_n$.

k variables, m equations, with m > k

The k-dimensional variant of Newton's method can be used to solve systems of greater than k (nonlinear) equations as well if the algorithm uses the generalized inverse of the non-square Jacobian matrix $J^+ = (J^T J)^{-1} J^T$ instead of the inverse of J. If the nonlinear system has no solution, the method attempts to find a solution in the non-linear least squares sense.

Nonlinear Equations in a Banach Space

Another generalization is Newton's method to find a root of a functional F defined in a Banach space. In this case the formulation is

$$X_{n+1} = X_n - \left(F'(X_n) \right)^{-1} F(X_n),$$

where $F'(X_n)$ is the Fréchet derivative computed at X_n. One needs the Fréchet derivative to be boundedly invertible at each X_n in order for the method to be applicable. A condition for existence of and convergence to a root is given by the Newton–Kantorovich theorem.

Nonlinear Equations over p-adic Numbers

In p-adic analysis, the standard method to show a polynomial equation in one variable has a p-adic root is Hensel's lemma, which uses the recursion from Newton's method on the p-adic numbers. Because of the more stable behavior of addition and multiplication in the p-adic numbers compared to the real numbers (specifically, the unit ball in the p-adics is a ring), convergence in Hensel's lemma can be guaranteed under much simpler hypotheses than in the classical Newton's method on the real line.

Newton–Fourier Method

The Newton–Fourier method is Joseph Fourier's extension of Newton's method to provide bounds on the absolute error of the root approximation, while still providing quadratic convergence.

Assume that $f(x)$ is twice continuously differentiable on $[a, b]$ and that f contains a root in this interval. Assume that $f'(x), f''(x) \neq 0$ on this interval (this is the case for instance if $f(a) < 0, f(b) > 0$, and $f'(x) > 0$, and $f''(x) > 0$ on this interval). This guarantees that there is a unique root on this interval, call it α. If it is concave down instead of concave up then replace $f(x)$ by $-f(x)$ since they have the same roots.

Let $x_0 = b$ be the right endpoint of the interval and let $z_0 = a$ be the left endpoint of the interval. Given x_n, define

$$x_{n+1} = x_n - \frac{f(x_n)}{f'(x_n)},$$

which is just Newton's method as before. Then define

$$z_{n+1} = z_n - \frac{f(z_n)}{f'(x_n)},$$

where the denominator is $f'(x_n)$ and not $f'(z_n)$. The iterations x_n will be strictly decreasing to the root while the iterations z_n will be strictly increasing to the root. Also,

$$\lim_{n \to \infty} \frac{x_{n+1} - z_{n+1}}{(x_n - z_n)^2} = \frac{f''(\alpha)}{2f'(\alpha)}$$

so that distance between x_n and z_n decreases quadratically.

Quasi-Newton Methods

When the Jacobian is unavailable or too expensive to compute at every iteration, a quasi-Newton method can be used.

Applications

Minimization and Maximization Problems

Newton's method can be used to find a minimum or maximum of a function. The derivative is zero at a minimum or maximum, so minima and maxima can be found by applying Newton's method to the derivative. The iteration becomes:

$$x_{n+1} = x_n - \frac{f'(x_n)}{f''(x_n)}.$$

Multiplicative Inverses of Numbers and Power Series

An important application is Newton–Raphson division, which can be used to quickly find the reciprocal of a number, using only multiplication and subtraction.

Finding the reciprocal of a amounts to finding the root of the function

$$f(x) = a - \frac{1}{x}$$

Newton's iteration is

$$x_{n+1} = x_n - \frac{f(x_n)}{f'(x_n)} = x_n - \frac{a - \dfrac{1}{x_n}}{\dfrac{1}{x_n^2}} = x_n(2 - ax_n)$$

Therefore, Newton's iteration needs only two multiplications and one subtraction.

This method is also very efficient to compute the multiplicative inverse of a power series.

Solving Transcendental Equations

Many transcendental equations can be solved using Newton's method. Given the equation

$$g(x) = h(x),$$

with $g(x)$ and/or $h(x)$ a transcendental function, one writes

$$f(x) = g(x) - h(x).$$

The values of x that solves the original equation are then the roots of $f(x)$, which may be found via Newton's method.

Examples

Square Root of a Number

Consider the problem of finding the square root of a number. Newton's method is one of many methods of computing square roots.

For example, if one wishes to find the square root of 612, this is equivalent to finding the solution to

$$x^2 = 612$$

The function to use in Newton's method is then

$$f(x) = x^2 - 612$$

with derivative

$$f'(x) = 2x.$$

With an initial guess of 10, the sequence given by Newton's method is

$$x_1 = x_0 - \frac{f(x_0)}{f'(x_0)} = 10 - \frac{10^2 - 612}{2 \times 10} = 35.6$$

$$x_2 = x_1 - \frac{f(x_1)}{f'(x_1)} = 35.6 - \frac{35.6^2 - 612}{2 \times 35.6} = \underline{26.}395505617978\ldots$$

$$x_3 = \quad \vdots \quad = \quad \vdots \quad = \underline{24.7}90635492455\ldots$$

$$x_4 = \quad \vdots \quad = \quad \vdots \quad = \underline{24.73}8688294075\ldots$$

$$x_5 = \quad \vdots \quad = \quad \vdots \quad = \underline{24.73863}3753767\ldots$$

where the correct digits are underlined. With only a few iterations one can obtain a solution accurate to many decimal places.

Solution of $\cos x = x^3$

Consider the problem of finding the positive number x with $\cos x = x^3$. We can re-phrase that as finding the zero of $f(x) = \cos(x) - x^3$. We have $f'(x) = -\sin(x) - 3x^2$. Since $\cos(x) \leq 1$ for all x and $x^3 > 1$ for $x > 1$, we know that our solution lies between 0 and 1. We try a starting value of $x_0 = 0.5$. (Note that a starting value of 0 will lead to an undefined result, showing the importance of using a starting point that is close to the solution.)

$$x_1 = x_0 - \frac{f(x_0)}{f'(x_0)} = 0.5 - \frac{\cos 0.5 - 0.5^3}{-\sin 0.5 - 3 \times 0.5^2} = 1.112141637097\ldots$$

$$x_2 = x_1 - \frac{f(x_1)}{f'(x_1)} = \vdots = \vdots = 0.909672693736\ldots$$

$$x_3 = \vdots = \vdots = \vdots = 0.867263818209\ldots$$

$$x_4 = \vdots = \vdots = \vdots = 0.865477135298\ldots$$

$$x_5 = \vdots = \vdots = \vdots = 0.8654740331 11\ldots$$

$$x_6 = \vdots = \vdots = \vdots = 0.865474033102\ldots$$

The correct digits are underlined in the above example. In particular, x_6 is correct to 12 decimal places. We see that the number of correct digits after the decimal point increases from 2 (for x_3) to 5 and 10, illustrating the quadratic convergence.

Pseudocode

The following is an example of using the Newton's Method to help find a root of a function f which has derivative fprime.

The initial guess will be $x_0 = 1$ and the function will be $f(x) = x^2 - 2$ so that $f'(x) = 2x$.

Each new iterative of Newton's method will be denoted by x1. We will check during the computation whether the denominator (yprime) becomes too small (smaller than epsilon), which would be the case if $f'(x_n) \approx 0$, since otherwise a large amount of error could be introduced.

```
%These choices depend on the problem being solved

x0 = 1 %The initial value

f = @(x) x^2 - 2 %The function whose root we are trying to find

fprime = @(x) 2*x %The derivative of f(x)

tolerance = 10^(-7) %7 digit accuracy is desired

epsilon = 10^(-14) %Don't want to divide by a number smaller than this

maxIterations = 20 %Don't allow the iterations to continue indefinitely

haveWeFoundSolution = false %Have not converged to a solution yet

for i = 1 : maxIterations

 y = f(x0)

 yprime = fprime(x0)

 if(abs(yprime) < epsilon) %Don't want to divide by too small of a number
```

```
% denominator is too small

break; %Leave the loop

end

x1 = x0 - y/yprime %Do Newton's computation

if(abs(x1 - x0) <= tolerance * abs(x1)) %If the result is within the
desired tolerance

haveWeFoundSolution = true

break; %Done, so leave the loop

end

x0 = x1 %Update x0 to start the process again

end

if (haveWeFoundSolution)

... % x1 is a solution within tolerance and maximum number of itera-
tions

else

... % did not converge

end
```

Analysis and Modeling of Evaporators Using Newton Raphson's Method

The Newton raphson method consist of the repeated use of linear terms of Taylor series expansions of the functions $f_1, f_2, f_3, f_4, f_5, f_6$.

$$0 = f_j + \frac{\partial f_j}{\partial V} \Delta Vo + \frac{\partial f_j}{\partial T} \Delta T_1 + \frac{\partial f_j}{\partial L} \Delta L_1 + \frac{\partial f_j}{\partial T} \Delta T_2 + \frac{\partial f_j}{\partial L} \Delta L_2 + \frac{\partial f_j}{\partial A} \Delta A$$

Where (j = 1,2,3,4......6)

$$\Delta V_0 = V_{k+1} - V_k;$$

$$\Delta T_1 = T_{k+1} - T_k \;;$$

$$\Delta T_2 = T_{2,k+1} - T_{2,k}$$

$$\Delta L_2 = L_{2,k+1} - L_k;$$

$$\Delta A = A_{K+1} - A_k;$$

Where the subscripts k and k+1 denotes the k th and k + I st trials

These six equations may be stated in compact form by means of the following matrix equation

$$J_k \Delta X_k = -f_k$$

Where J_k is called Jacobin matrix and

$$\Delta X_k = X_{k+1} - X_k = [\Delta V_0 \Delta T_1 \Delta T_2 \Delta L_2 \Delta A]^T$$

The subscripts k and k+1 denotes that the elements of the matrices carrying these subscripts are those given by k and k+1st trials, respectively. In the interest of simplicity k is omitted from the elements of X_k, J_k, f_k. On the basis of set assumed values for the elements of column vector X, which may be stated as transpose of the corresponding row vector.

$$X_k = [V_0 T_1 L_1 T_2 L_2 A]$$

The corresponding values of elements of J and f are computed.

A display of the elements of J_k and f_k follows-:

J_k

$$\begin{vmatrix} \partial f_1/\partial V & \partial f_1/\partial T & \partial f_1/\partial L & \partial f_1/\partial T & \partial f_1/\partial L & \partial f_1/\partial A \\ \partial f_2/\partial V & \partial f_2/\partial T & \partial f_2/\partial L & \partial f_2/\partial T & \partial f_2/\partial L & \partial f_2/\partial A \\ - & - & - & - & - & - \\ - & - & - & - & - & - \\ - & - & - & - & - & - \\ - & - & - & - & - & - \\ \partial f_6/\partial V & \partial f_6/\partial T & \partial f_6/\partial L & \partial f_6/\partial T & \partial f_6/\partial L & \partial f_6/\partial A \end{vmatrix}$$

$$f_k = [f_1, f_2, f_3, f_4, f_5, f_6]^T$$

The functions $f_1, f_2, f_3, f_4, f_5, f_6$ and their partial derivative which appear in J_k are continuous and the determinant of j is not equal to zero, then only Newton raphson method will converge, provided a set of assumed values of the variables which are closed enough to the solution can be found.

Let us consider a triple effect evaporator with forward feed arrangement without boiling point elevation. The functions $f_1, f_2, f_3, f_4, f_5, f_6$ obtained are stated below-

1 Effect-

Enthalpy balance-

$$f_1 = F(h_f - h_1) + V_0 \lambda_0 - (F - L_1)\lambda_1$$

Heat transfer rate-

$$f_2 = UA(T_0 - T) - V_0 \lambda_0$$

2 Effect-

Enthalpy balance-

$$f_3 = L_1(h_1 - h_2) + (F - L)\lambda_1 - (L_1 - L_2)\lambda_2$$

Heat transfer rate-

$$f_4 = U_2 A(T_1 - T_2) - (F - L_1)\lambda_1$$

3 Effect-

Enthalpy balance-

$$f_5 = L_2 (h_2 - h_3) + (L_1 - L_2)\lambda_2 - (L_2 - L_3)\lambda_3$$

Heat transfer rate-

$$f_6 = U_3 A(T_2 - T_3) - (L_1 - L_2)\lambda_2$$

If the changes in the specific heat with temperature in the neighbourhood of the solution to equations are negligible, then the sensible heat terms $(h_F\text{-}h_1),(h_1\text{-}h_2),(h_2\text{-}h_3)$ may be replaced by the respective equivalents $C_p (T_F\text{-}T_1), C_p(T_1\text{-}T_2)$ and $C (T_2\text{-}T_3)$. If the variation of the latent heats with temperature

Are also regarded as negligible in the neighbourhood of the solution, then

$J_k =$

$$\begin{vmatrix} -\lambda_0 & -FC_p & \lambda_1 & 0 & 0 & 0 \\ -\lambda_0 & -U_1 A & 0 & 0 & 0 & U_1(T_0 - T_1) \\ 0 & L_1 C_p & b_{33} & -L_1 C_p & \lambda_2 & 0 \\ 0 & U_2 A & \lambda_1 & -U_2 A & 0 & U_2(T_1 - T_2) \\ 0 & 0 & -\lambda_2 & L_2 C_p & b_{55} & 0 \\ 0 & 0 & -\lambda_2 & U_3 A & \lambda_2 & U_3(T_2 - T_3) \end{vmatrix}$$

Where $b_{33} = C_p(T_1 - T_2) - (\lambda_1 + \lambda_2)$

$$b_{55} = C_p(T_2 - T_3) - (\lambda_2 + \lambda_3)$$

Example:

It is desired to design a triple effect evaporator system to concentrate the solute from a 10% solution to 50% by wt. The feed rate is 50,000 lb/hr and it enters the first effect liquid as liquid at 100 °F. Forward feed is to be used. Saturated vapour of the solvent at 250°F is available for satisfying the heating requirement for the first effect. The third effect is to be operated at an absolute pressure corresponding to a boiling point of 125°F for the pure solvent. Neglect boiling point elevation as well as the variation of heat capacities and latent heat of vaporisation with temperature and composition. Determine the area per effect the temperature T_1 and T_2 and the flow rates and the composition and the rate.

$$Given - C_p = 1 btu \, / \, lb° \, F$$
$$\lambda_{20} = \lambda_1 = \lambda_3 = 1000 \; btu \, / \, lb° \, F$$
$$U_1 = 500, U_2 = 300, U_3 = 200$$

Solution:

The calculation procedure may be based initatedon the basis of reasonable set of assumptions,

1. T_0-T_1 =42°F, T_1-T_2 =42°F, T_2-T_3 =41°F

2. Solvent evaporated in first effect =14,000lb/hr

Solvent evaporated in second effect =14,000lb/hr

Solvent evaporated in third effect =12,000 lb/hr

3. A =1000 ft for each effect

4. V_0 = 15000/hr

A scaling procedure is used to reduce to reduce the magnitude of the terms appearing in the functional group and matrices. For computational purposes, it is desirable to have tems with magnitude near unity. The following scaling procedure easy used.

1. Each functional group is divided by the product $F\lambda_0$ and new functional group so obtained is denoted by g_j.

2. All flow rates were expressed as afraction of the feed rate F that is $L_j = l_j F$ and $V_j = v_j F$

3. All temperature temperatures were expressed as function of steam temperature as follow:

$$T_j = u_j T_0$$

4. The area of each effect is expressed as a fration of term proportional to the feed rate in the following manner: $A_j = a_j(F/50)$

After this scaling procedure has been applied to the functional expressions, the matrices J, ΔX

$$\Delta X_k = [\Delta v_0 \ \Delta u_1 \ \Delta l_1 \ \Delta u_2 \ \Delta l_2 \ \Delta a]^T$$

$$f_k = [g_1 \ g_2 \ g_3 \ g_4 \ g_5 \ g_6]^T$$

$j_k =$

$$\begin{vmatrix}
-1 & b_{12} & \lambda_1/\lambda_0 & 0 & 0 & 0 \\
-1 & b_{22} & 0 & 0 & 0 & b_{26} \\
0 & b_{33} & b_{33} & b_{34} & \lambda_2/\lambda_0 & 0 \\
0 & b_{42} & \lambda_1/\lambda_0 & b_{44} & 0 & b_{45} \\
0 & 0 & -\lambda_2/\lambda & b_{54} & b_{55} & 0 \\
0 & 0 & -\lambda_2/\lambda & b_{64} & \lambda_2/\lambda_0 & b_{65}
\end{vmatrix}$$

Where the elements of J consist of the partial derivative of the g's with respect to new set of variables.

$b_{12} = -C_p T_0/\lambda_0; b_{34} = -l_1 C_p T_0/\lambda_0; b_{22} = -U_1 a T_0/50\lambda_0; b_{44} = U_2 a T_0/50\lambda_0;; b_{32} = -l_1 C_p T_0/\lambda_0;$
$b_{54} = -l_2 C_p T_0/\lambda_0; b_{42} = -U_2 a T_0/50\lambda_0; b_{64} = -U_2 a T_0/50\lambda_0; b_{26} = -[U_1(1-u_1)T_0]/50\lambda_0;$
$b_{33} = [C_p(u_1-u_2)T_0 - (\lambda_1+\lambda_2)]/\lambda_0; b_{46} = [U_2(u_1-u_2)T_0]/50\lambda_0]; b_{66} = [U_3(u_2-u_3)T_0]/50\lambda_0;$
$b_{55} = [C_p(u_2-u_3)T_0 - (\lambda_2+\lambda_3)]/\lambda_0$

Then on the basis of the assumed values of the variables, the matrices J_0, f_0 and X_0 are used in the first trial follow-:

$J_0 =$

$$\begin{vmatrix}
1 & -2.5 & 1 & 0 & 0 & 0 \\
-1 & -2.5 & 0 & 0 & 0 & .42 \\
0 & .18 & -1.958 & -.18 & 1 & 0 \\
0 & 1.5 & 1 & -1.5 & 0 & .252 \\
0 & 0 & 1 & .11 & -1.959 & 0 \\
0 & 0 & -1 & 1 & 1 & .614
\end{vmatrix}$$

$$f_0 = [-.088.120.030 - .028.058 - .116]^T$$

$$X_0 = [.300.832.720.664.4401.000]^T$$

Convergence to within about six significant numbers was achieved in three trials. The values of X complicated at the end of first four trials are displayed.

Table: Calculate Value of Unknown Variables

Trial no	vo	u_1	l_1	u_2	l_2	A
1	.359374	.879493	.760498	.743069	.494740	1.13835
2	.357773	.874144	.760762	.733878	.494848	1.13703
3	.357771	.874138	.760763	.733868	.494848	1.13703
4	.357771	.874138	.760763	.733868	.494848	1.13703

Thus the desired solution is-:

$$V_0 = v_0 F = 17,888.5 lb / hr; T_1 = u_1 T_0 = 183.467 F; L_1 = l_1 F = 38,038.1 lb / hr; L_2 - l_2 F = 24,742.4 lb / hr;$$
$$A - (aF/50) = 1,137.03 lb / hr; L_3 = FX / x_3 = 10,000 lb / hr; x_1 = FX / L_1 = .131447; X = FX / L_2 = .202082$$

Distillation

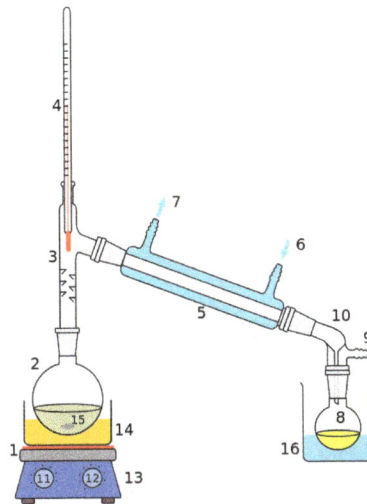

Laboratory display of distillation: 1: A source of heat 2: Still pot 3: Still head 4: Thermometer/Boiling point temperature 5: Condenser 6: Cooling water in 7: Cooling water out 8: Distillate/receiving flask 9: Vacuum/gas inlet 10: Still receiver 11: Heat control 12: Stirrer speed control 13: Stirrer/heat plate 14: Heating (Oil/sand) bath 15: Stirring means e.g. (shown), boiling chips or mechanical stirrer 16: Cooling bath

Distillation is the process of separating the component or substances from a liquid mixture by selective evaporation and condensation. Distillation may result in essentially complete separation (nearly pure components), or it may be a partial separation that increases the concentration of selected components of the mixture. In either case the process exploits differences in the volatility of the mixture's components. In industrial chemistry, distillation is a unit operation of practically universal importance, but it is a physical separation process and not a chemical reaction.

Commercially, distillation has many applications. For example:

- In the fossil fuel industry distillation is a major class of operation in obtaining materials from crude oil for fuels and for chemical feed stocks.

- Distillation permits separation of air into its components — notably oxygen, nitrogen, and argon — for industrial use.

- In the field of industrial chemistry, large amounts of crude liquid products of chemical synthesis are distilled to separate them, either from other products, or from impurities, or from unreacted starting materials.

- Distillation of fermented products produces distilled beverages with a high alcohol content, or separates out other fermentation products of commercial value.

An installation for distillation, especially of alcohol, is a distillery. The distillation equipment is a still.

Distillation is a very old method of artificial desalination.

History

Distillation equipment used by the 3rd century Greek alchemist Zosimos of Panopolis, from the Byzantine Greek manuscript *Parisinus graces*

Aristotle wrote about the process in his *Meteorologica* and even that "ordinary wine possesses a kind of exhalation, and that is why it gives out a flame". Later evidence of distillation comes from Greek alchemists working in Alexandria in the 1st century AD. Distilled water has been known since at least c. 200, when Alexander of Aphrodisias described the process. Work on distilling other liquids continued in early Byzantine Egypt under the Greek-Egyptian Zosimus of Panopolis. Distillation in China could have begun during the Eastern Han Dynasty (1st–2nd centuries), but archaeological evidence indicates that actual distillation of beverages began in the Jin (12th–13th centuries) and Southern Song (10th–13th centuries) dynasties. A still was found in an archaeological site in Qinglong, Hebei province dating to the 12th century. Distilled beverages were more common during the Yuan dynasty (13th–14th centuries). Arabs learned the process from the Alexandrians and used it extensively in their chemical experiments.

Clear evidence of the distillation of alcohol comes from the School of Salerno in the 12th century. Fractional distillation was developed by Tadeo Alderotti in the 13th century.

In 1500, German alchemist Hieronymus Braunschweig published *Liber de arte destillandi* (The Book of the Art of Distillation) the first book solely dedicated to the subject of distillation, followed in 1512 by a much expanded version. In 1651, John French published The Art of Distillation the first major English compendium of practice, though it has been claimed that much of it derives from Braunschweig's work. This includes diagrams with people in them showing the industrial rather than bench scale of the operation.

Hieronymus Brunschwig's *Liber de arte Distillandi de Compositis* (Strassburg, 1512) Chemical Heritage Foundation

A retort

As alchemy evolved into the science of chemistry, vessels called retorts became used for distillations. Both alembics and retorts are forms of glassware with long necks pointing to the side at a downward angle which acted as air-cooled condensers to condense the distillate and let it drip downward for collection. Later, copper alembics were invented. Riveted joints were often kept tight by using various mixtures, for instance a dough made of rye flour. These alembics often featured a cooling system around the beak, using cold water for instance, which made the condensation of alcohol more efficient. These were called pot stills. Today, the retorts and pot stills have been largely supplanted by more efficient distillation methods in most industrial processes. However, the pot still is still widely used for the elaboration of some fine alcohols such as cognac, Scotch whisky, tequila and some vodkas. Pot stills made of various materials (wood, clay, stainless steel) are also used by bootleggers in various countries. Small pot stills are also sold for the domestic production of flower water or essential oils.

Distillation

Early forms of distillation were batch processes using one vaporization and one condensation. Purity was improved by further distillation of the condensate. Greater volumes were processed by simply repeating the distillation. Chemists were reported to carry out as many as 500 to 600 distillations in order to obtain a pure compound.

Old Ukrainian vodka still

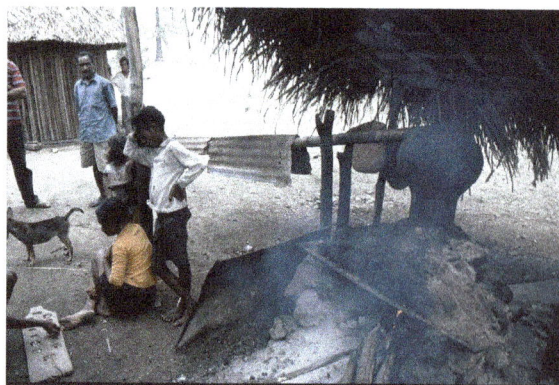

Simple liqueur distillation in East Timor

In the early 19th century the basics of modern techniques including pre-heating and reflux were developed. In 1822, Anthony Perrier developed one of the first continuous stills. In 1826, Robert Stein improved that design to make his patent still. In 1830, Aeneas Coffey got a patent for improving that design. Coffey's continuous still may be regarded as the archetype of modern petrochemical units. The French engineer Armand Savalle developed his steam regulator around 1846. In 1877, Ernest Solvay was granted a U.S. Patent for a tray column for ammonia distillation and the same and subsequent years saw developments of this theme for oil and spirits.

With the emergence of chemical engineering as a discipline at the end of the 19th century, scientific rather than empirical methods could be applied. The developing petroleum industry in the early 20th century provided the impetus for the development of accurate design methods such as the McCabe–Thiele method and the Fenske equation. The availability of powerful computers has also allowed direct computer simulation of distillation columns.

Applications of Distillation

The application of distillation can roughly be divided in four groups: laboratory scale, industrial distillation, distillation of herbs for perfumery and medicinals (herbal distillate), and food processing. The latter two are distinctively different from the former two in that in the processing of beverages and herbs, the distillation is not used as a true purification method but more to transfer all volatiles from the source materials to the distillate.

The main difference between laboratory scale distillation and industrial distillation is that laboratory scale distillation is often performed batch-wise, whereas industrial distillation often occurs continuously. In batch distillation, the composition of the source material, the vapors of the distilling compounds and the distillate change during the distillation. In batch distillation, a still is charged (supplied) with a batch of feed mixture, which is then separated into its component fractions which are collected sequentially from most volatile to less volatile, with the bottoms (remaining least or non-volatile fraction) removed at the end. The still can then be recharged and the process repeated.

In continuous distillation, the source materials, vapors, and distillate are kept at a constant composition by carefully replenishing the source material and removing fractions from both vapor and liquid in the system. This results in a better control of the separation process.

Idealized Distillation Model

The boiling point of a liquid is the temperature at which the vapor pressure of the liquid equals the pressure around the liquid, enabling bubbles to form without being crushed. A special case is the normal boiling point, where the vapor pressure of the liquid equals the ambient atmospheric pressure.

It is a common misconception that in a liquid mixture at a given pressure, each component boils at the boiling point corresponding to the given pressure and the vapors of each component will collect separately and purely. This, however, does not occur even in an idealized system. Idealized models of distillation are essentially governed by Raoult's law and Dalton's law, and assume that vapor–liquid equilibria are attained.

Raoult's law states that the vapor pressure of a solution is dependent on 1) the vapor pressure of each chemical component in the solution and 2) the fraction of solution each component makes up a.k.a. the mole fraction. This law applies to ideal solutions, or solutions that have different components but whose molecular interactions are the same as or very similar to pure solutions.

Dalton's law states that the total pressure is the sum of the partial pressures of each individual component in the mixture. When a multi-component liquid is heated, the vapor pressure of each component will rise, thus causing the total vapor pressure to rise. When the total vapor pressure reaches the pressure surrounding the liquid, boiling occurs and liquid turns to gas throughout the bulk of the liquid. Note that a mixture with a given composition has one boiling point at a given pressure, when the components are mutually soluble. A mixture of constant composition does not have multiple boiling points.

An implication of one boiling point is that lighter components never cleanly "boil first". At boiling point, all volatile components boil, but for a component, its percentage in the vapor is the same as its percentage of the total vapor pressure. Lighter components have a higher partial pressure and thus are concentrated in the vapor, but heavier volatile components also have a (smaller) partial pressure and necessarily evaporate also, albeit being less concentrated in the vapor. Indeed, batch distillation and fractionation succeed by varying the composition of the mixture. In batch distillation, the batch evaporates, which changes its composition; in fractionation, liquid higher in the fractionation column contains more lights and boils at lower temperatures. Therefore, starting from a given mixture, it appears to have a boiling range instead of a boiling *point*, although this is because its composition changes: each intermediate mixture has its own, singular boiling point.

The idealized model is accurate in the case of chemically similar liquids, such as benzene and toluene. In other cases, severe deviations from Raoult's law and Dalton's law are observed, most famously in the mixture of ethanol and water. These compounds, when heated together, form an azeotrope, which is a composition with a boiling point higher or lower than the boiling point of each separate liquid. Virtually all liquids, when mixed and heated, will display azeotropic behaviour. Although there are computational methods that can be used to estimate the behavior of a mixture of arbitrary components, the only way to obtain accurate vapor–liquid equilibrium data is by measurement.

It is not possible to *completely* purify a mixture of components by distillation, as this would require each component in the mixture to have a zero partial pressure. If ultra-pure products are the goal, then further chemical separation must be applied. When a binary mixture is evaporated and the other component, e.g. a salt, has zero partial pressure for practical purposes, the process is simpler and is called evaporation in engineering.

Batch Distillation

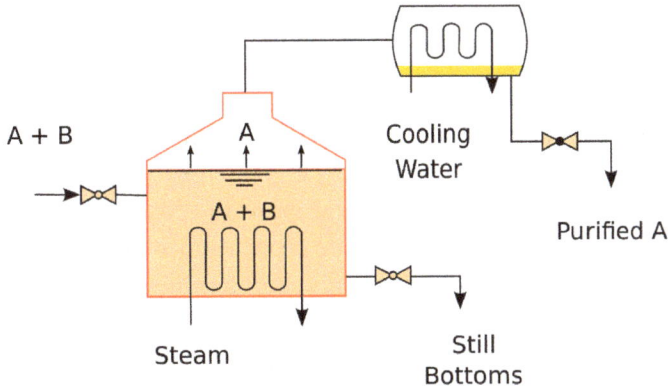

A batch still showing the separation of A and B.

Heating an ideal mixture of two volatile substances A and B (with A having the higher volatility, or lower boiling point) in a batch distillation setup (such as in an apparatus depicted in the opening figure) until the mixture is boiling results in a vapor above the liquid which contains a mixture of A and B. The ratio between A and B in the vapor will be different from the ratio in the liquid: the ratio in the liquid will be determined by how the original mixture was prepared, while the ratio in the vapor will be enriched in the more volatile compound, A. The vapor goes through the condenser and is removed from the system. This in turn means that the ratio of compounds in the remaining liquid is now different from the initial ratio (i.e., more enriched in B than the starting liquid).

The result is that the ratio in the liquid mixture is changing, becoming richer in component B. This causes the boiling point of the mixture to rise, which in turn results in a rise in the temperature in the vapor, which results in a changing ratio of A : B in the gas phase (as distillation continues, there is an increasing proportion of B in the gas phase). This results in a slowly changing ratio A : B in the distillate.

If the difference in vapor pressure between the two components A and B is large (generally expressed as the difference in boiling points), the mixture in the beginning of the distillation is highly enriched in component A, and when component A has distilled off, the boiling liquid is enriched in component B.

Continuous Distillation

Continuous distillation is an ongoing distillation in which a liquid mixture is continuously (without interruption) fed into the process and separated fractions are removed continuously as output streams occur over time during the operation. Continuous distillation produces a minimum of two output fractions, including at least one volatile distillate fraction, which has boiled and been separately captured as a vapor, and then condensed to a liquid. There is always a bottoms (or residue) fraction, which is the least volatile residue that has not been separately captured as a condensed vapor.

Continuous distillation differs from batch distillation in the respect that concentrations should not change over time. Continuous distillation can be run at a steady state for an arbitrary amount of time. For any source material of specific composition, the main variables that affect the purity of products in continuous distillation are the reflux ratio and the number of theoretical equilibrium stages, in practice determined by the number of trays or the height of packing. Reflux is a flow from the condenser back to the column, which generates a recycle that allows a better separation with a given number of trays. Equilibrium stages are ideal steps where compositions achieve vapor–liquid equilibrium, repeating the separation process and allowing better separation given a reflux ratio. A column with a high reflux ratio may have fewer stages, but it refluxes a large amount of liquid, giving a wide column with a large holdup. Conversely, a column with a low reflux ratio must have a large number of stages, thus requiring a taller column.

General Improvements

Both batch and continuous distillations can be improved by making use of a fractionating column on top of the distillation flask. The column improves separation by providing a larger surface area for the vapor and condensate to come into contact. This helps it remain at equilibrium for as long as possible. The column can even consist of small subsystems ('trays' or 'dishes') which all contain an enriched, boiling liquid mixture, all with their own vapor–liquid equilibrium.

There are differences between laboratory-scale and industrial-scale fractionating columns, but the principles are the same. Examples of laboratory-scale fractionating columns (in increasing efficiency) include

- Air condenser

- Vigreux column (usually laboratory scale only)

- Packed column (packed with glass beads, metal pieces, or other chemically inert material)

- Spinning band distillation system.

Laboratory Scale Distillation

Laboratory scale distillations are almost exclusively run as batch distillations. The device used in distillation, sometimes referred to as a *still*, consists at a minimum of a reboiler or pot in which the source material is heated, a condenser in which the heated vapour is cooled back to the liquid state, and a receiver in which the concentrated or purified liquid, called the distillate, is collected. Several laboratory scale techniques for distillation exist.

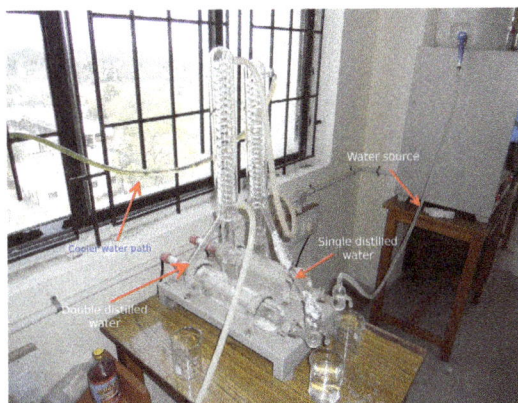

Typical laboratory distillation unit

Simple Distillation

In simple distillation, the vapor is immediately channeled into a condenser. Consequently, the distillate is not pure but rather its composition is identical to the composition of the vapors at the given temperature and pressure. That concentration follows Raoult's law.

As a result, simple distillation is effective only when the liquid boiling points differ greatly (rule of thumb is 25 °C) or when separating liquids from non-volatile solids or oils. For these cases, the vapor pressures of the components are usually different enough that the distillate may be sufficiently pure for its intended purpose.

Fractional Distillation

For many cases, the boiling points of the components in the mixture will be sufficiently close that Raoult's law must be taken into consideration. Therefore, fractional distillation must be used in order to separate the components by repeated vaporization-condensation cycles within a packed fractionating column. This separation, by successive distillations, is also referred to as rectification.

As the solution to be purified is heated, its vapors rise to the fractionating column. As it rises, it cools, condensing on the condenser walls and the surfaces of the packing material. Here, the condensate continues to be heated by the rising hot vapors; it vaporizes once more. However, the composition of the fresh vapors are determined once again by Raoult's law. Each vaporization-condensation cycle (called a *theoretical plate*) will yield a purer solution of the more volatile component. In reality, each cycle at a given temperature does not occur at exactly the same position in the fractionating column; *theoretical plate* is thus a concept rather than an accurate description.

More theoretical plates lead to better separations. A spinning band distillation system uses a spinning band of Teflon or metal to force the rising vapors into close contact with the descending condensate, increasing the number of theoretical plates.

Steam Distillation

Dimethyl sulfoxide usually boils at 189 °C. Under a vacuum, it distills off into the receiver at only 70 °C.

Like vacuum distillation, steam distillation is a method for distilling compounds which are heat-sensitive. The temperature of the steam is easier to control than the surface of a heating element, and allows a high rate of heat transfer without heating at a very high temperature. This process involves bubbling steam through a heated mixture of the raw material. By Raoult's law, some of the target compound will vaporize (in accordance with its partial pressure). The vapor mixture is cooled and condensed, usually yielding a layer of oil and a layer of water.

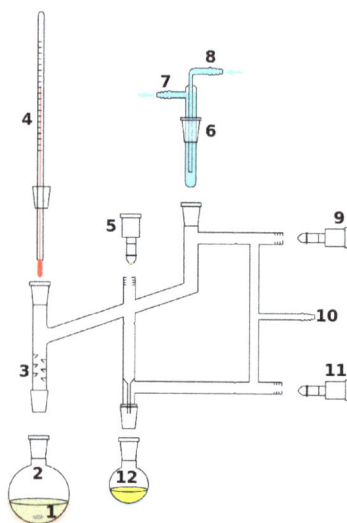

Perkin triangle distillation setup
1: Stirrer bar/anti-bumping granules 2: Still pot 3: Fractionating column 4: Thermometer/Boiling point temperature 5: Teflon tap 1 6: Cold finger 7: Cooling water out 8: Cooling water in 9: Teflon tap 2 10: Vacuum/gas inlet 11: Teflon tap 3 12: Still receiver

Steam distillation of various aromatic herbs and flowers can result in two products; an essential oil as well as a watery herbal distillate. The essential oils are often used in perfumery and aromatherapy while the watery distillates have many applications in aromatherapy, food processing and skin care.

Vacuum Distillation

Some compounds have very high boiling points. To boil such compounds, it is often better to lower the pressure at which such compounds are boiled instead of increasing the temperature. Once the pressure is lowered to the vapor pressure of the compound (at the given temperature), boiling and the rest of the distillation process can commence. This technique is referred to as vacuum distillation and it is commonly found in the laboratory in the form of the rotary evaporator.

This technique is also very useful for compounds which boil beyond their decomposition temperature at atmospheric pressure and which would therefore be decomposed by any attempt to boil them under atmospheric pressure.

Molecular distillation is vacuum distillation below the pressure of 0.01 torr. 0.01 torr is one order of magnitude above high vacuum, where fluids are in the free molecular flow regime, i.e. the mean free path of molecules is comparable to the size of the equipment. The gaseous phase no longer exerts significant pressure on the substance to be evaporated, and consequently, rate of evaporation no longer depends on pressure. That is, because the continuum assumptions of fluid dynamics no longer apply, mass transport is governed by molecular dynamics rather than fluid dynamics. Thus, a short path between the hot surface and the cold surface is necessary, typically by suspending a hot plate covered with a film of feed next to a cold plate with a line of sight in between. Molecular distillation is used industrially for purification of oils.

Air-sensitive Vacuum Distillation

Some compounds have high boiling points as well as being air sensitive. A simple vacuum distillation system as exemplified above can be used, whereby the vacuum is replaced with an inert gas after the distillation is complete. However, this is a less satisfactory system if one desires to collect fractions under a reduced pressure. To do this a "cow" or "pig" adaptor can be added to the end of the condenser, or for better results or for very air sensitive compounds a Perkin triangle apparatus can be used.

The Perkin triangle, has means via a series of glass or Teflon taps to allows fractions to be isolated from the rest of the still, without the main body of the distillation being removed from either the vacuum or heat source, and thus can remain in a state of reflux. To do this, the sample is first isolated from the vacuum by means of the taps, the vacuum over the sample is then replaced with an inert gas (such as nitrogen or argon) and can then be stoppered and removed. A fresh collection vessel can then be added to the system, evacuated and linked back into the distillation system via the taps to collect a second fraction, and so on, until all fractions have been collected.

Short Path Distillation

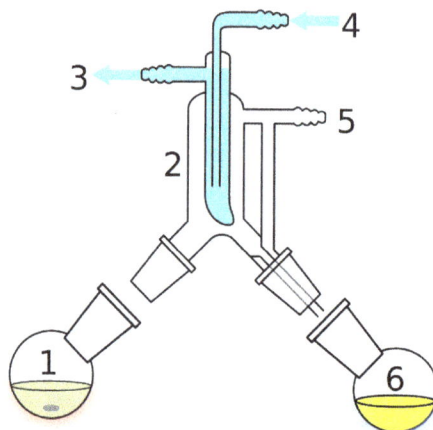

Short path vacuum distillation apparatus with vertical condenser (cold finger), to minimize the distillation path; 1: Still pot with stirrer bar/anti-bumping granules 2: Cold finger – bent to direct condensate 3: Cooling water out 4: cooling water in **5:** Vacuum/gas inlet **6:** Distillate flask/distillate

Short path distillation is a distillation technique that involves the distillate travelling a short distance, often only a few centimeters, and is normally done at reduced pressure. A classic example would be a distillation involving the distillate travelling from one glass bulb to another, without the need for a condenser separating the two chambers. This technique is often used for compounds which are unstable at high temperatures or to purify small amounts of compound. The advantage is that the heating temperature can be considerably lower (at reduced pressure) than the boiling point of the liquid at standard pressure, and the distillate only has to travel a short distance before condensing. A short path ensures that little compound is lost on the sides of the apparatus. The Kugelrohr is a kind of a short path distillation apparatus which often contain multiple chambers to collect distillate fractions.

Zone Distillation

Zone distillation is a distillation process in long container with partial melting of refined matter in moving liquid zone and condensation of vapor in the solid phase at condensate pulling in cold area. The process is worked in theory. When zone heater is moving from the top to the bottom of the container then solid condensate with irregular impurity distribution is forming. Then most pure part of the condensate may be extracted as product. The process may be iterated many times by moving (without turnover) the received condensate to the bottom part of the container on the place of refined matter. The irregular impurity distribution in the condensate (that is efficiency of purification) increases with number of repetitions of the process. Zone distillation is a distillation analog of zone recrystallization. Impurity distribution in the condensate is described by known equations of zone recrystallization with various numbers of iteration of process – with replacement distribution efficient k of crystallization on separation factor α of distillation.

Other Types

- The process of reactive distillation involves using the reaction vessel as the still. In this process, the product is usually significantly lower-boiling than its reactants. As the product is formed from the reactants, it is vaporized and removed from the reaction mixture. This technique is an example of a continuous vs. a batch process; advantages include less downtime to charge the reaction vessel with starting material, and less workup. Distillation "over a reactant" could be classified as a reactive distillation. It is typically used to remove volatile impurity from the distallation feed. For example, a little lime may be added to remove carbon dioxide from water followed by a second distillation with a little sulfuric acid added to remove traces of ammonia.

- Catalytic distillation is the process by which the reactants are catalyzed while being distilled to continuously separate the products from the reactants. This method is used to assist equilibrium reactions reach completion.

- Pervaporation is a method for the separation of mixtures of liquids by partial vaporization through a non-porous membrane.

- Extractive distillation is defined as distillation in the presence of a miscible, high boiling, relatively non-volatile component, the solvent, that forms no azeotrope with the other components in the mixture.

- Flash evaporation (or partial evaporation) is the partial vaporization that occurs when a saturated liquid stream undergoes a reduction in pressure by passing through a throttling valve or other throttling device. This process is one of the simplest unit operations, being equivalent to a distillation with only one equilibrium stage.

- Codistillation is distillation which is performed on mixtures in which the two compounds are not miscible. In the laboratory, the Dean-Stark apparatus is used for this purpose to remove water from synthesis products.

The unit process of evaporation may also be called "distillation":

- In rotary evaporation a vacuum distillation apparatus is used to remove bulk solvents from a sample. Typically the vacuum is generated by a water aspirator or a membrane pump.

- In a kugelrohr a short path distillation apparatus is typically used (generally in combination with a (high) vacuum) to distill high boiling (> 300 °C) compounds.

The apparatus consists of an oven in which the compound to be distilled is placed, a receiving portion which is outside of the oven, and a means of rotating the sample. The vacuum is normally generated by using a high vacuum pump.

Other uses:

- Dry distillation or destructive distillation, despite the name, is not truly distillation, but rather a chemical reaction known as pyrolysis in which solid substances are heated in an inert or reducing atmosphere and any volatile fractions, containing high-boiling liquids and products of pyrolysis, are collected. The destructive distillation of wood to give methanol is the root of its common name – *wood alcohol*.

- Freeze distillation is an analogous method of purification using freezing instead of evaporation. It is not truly distillation, but a recrystallization where the product is the mother liquor, and does not produce products equivalent to distillation. This process is used in the production of ice beer and ice wine to increase ethanol and sugar content, respectively. It is also used to produce applejack. Unlike distillation, freeze distillation concentrates poisonous congeners rather than removing them; As a result, many countries prohibit such applejack as a health measure. However, reducing methanol with the absorption of 4A molecular sieve is a practical method for production. Also, distillation by evaporation can separate these since they have different boiling points.

Azeotropic Distillation

Interactions between the components of the solution create properties unique to the solution, as most processes entail nonideal mixtures, where Raoult's law does not hold. Such interactions can result in a constant-boiling azeotrope which behaves as if it were a pure compound (i.e., boils at a single temperature instead of a range). At an azeotrope, the solution contains the given component in the same proportion as the vapor, so that evaporation does not change the purity, and distillation does not effect separation. For example, ethyl alcohol and water form an azeotrope of 95.6% at 78.1 °C.

If the azeotrope is not considered sufficiently pure for use, there exist some techniques to break the azeotrope to give a pure distillate. This set of techniques are known as azeotropic distillation. Some techniques achieve this by "jumping" over the azeotropic composition (by adding another component to create a new azeotrope, or by varying the pressure). Others work by chemically or physically removing or sequestering the impurity. For example, to purify ethanol beyond 95%, a drying agent (or desiccant, such as potassium carbonate) can be added to convert the soluble water into insoluble water of crystallization. Molecular sieves are often used for this purpose as well.

Immiscible liquids, such as water and toluene, easily form azeotropes. Commonly, these azeotropes are referred to as a low boiling azeotrope because the boiling point of the azeotrope is lower than the boiling point of either pure component. The temperature and composition of the azeotrope is easily predicted from the vapor pressure of the pure components, without use of Raoult's law. The azeotrope is easily broken in a distillation set-up by using a liquid–liquid separator (a decanter) to separate the two liquid layers that are condensed overhead. Only one of the two liquid layers is refluxed to the distillation set-up.

High boiling azeotropes, such as a 20 weight percent mixture of hydrochloric acid in water, also exist. As implied by the name, the boiling point of the azeotrope is greater than the boiling point of either pure component.

To break azeotropic distillations and cross distillation boundaries, such as in the DeRosier Problem, it is necessary to increase the composition of the light key in the distillate.

Breaking an Azeotrope with Unidirectional Pressure Manipulation

The boiling points of components in an azeotrope overlap to form a band. By exposing an azeotrope to a vacuum or positive pressure, it's possible to bias the boiling point of one component away from the other by exploiting the differing vapour pressure curves of each; the curves may overlap at the azeotropic point, but are unlikely to be remain identical further along the pressure axis either side of the azeotropic point. When the bias is great enough, the two boiling points no longer overlap and so the azeotropic band disappears.

This method can remove the need to add other chemicals to a distillation, but it has two potential drawbacks.

Under negative pressure, power for a vacuum source is needed and the reduced boiling points of the distillates requires that the condenser be run cooler to prevent distillate vapours being lost to the vacuum source. Increased cooling demands will often require additional energy and possibly new equipment or a change of coolant.

Alternatively, if positive pressures are required, standard glassware can not be used, energy must be used for pressurization and there is a higher chance of side reactions occurring in the distillation, such as decomposition, due to the higher temperatures required to effect boiling.

A unidirectional distillation will rely on a pressure change in one direction, either positive or negative.

Pressure-swing Distillation

Pressure-swing distillation is essentially the same as the unidirectional distillation used to break azeotropic mixtures, but here both positive and negative pressures may be employed.

This improves the selectivity of the distillation and allows a chemist to optimize distillation by avoiding extremes of pressure and temperature that waste energy. This is particularly important in commercial applications.

One example of the application of pressure-swing distillation is during the industrial purification of ethyl acetate after its catalytic synthesis from ethanol.

Industrial Distillation

Typical industrial distillation towers

Large scale industrial distillation applications include both batch and continuous fractional, vacuum, azeotropic, extractive, and steam distillation. The most widely used industrial applications of continuous, steady-state fractional distillation are in petroleum refineries, petrochemical and chemical plants and natural gas processing plants.

To control and optimize such industrial distillation, a standardized laboratory method, ASTM D86, is established. This test method extends to the atmospheric distillation of petroleum products using a laboratory batch distillation unit to quantitatively determine the boiling range characteristics of petroleum products.

Industrial distillation is typically performed in large, vertical cylindrical columns known as distillation towers or distillation columns with diameters ranging from about 65 centimeters to 16 meters and heights ranging from about 6 meters to 90 meters or more. When the process feed has a diverse composition, as in distilling crude oil, liquid outlets at intervals up the column allow for the withdrawal of different fractions or products having different boiling points or boiling ranges. The "lightest" products (those with the lowest boiling point) exit from the top of the columns and the "heaviest" products (those with the highest boiling point) exit from the bottom of the column and are often called the bottoms.

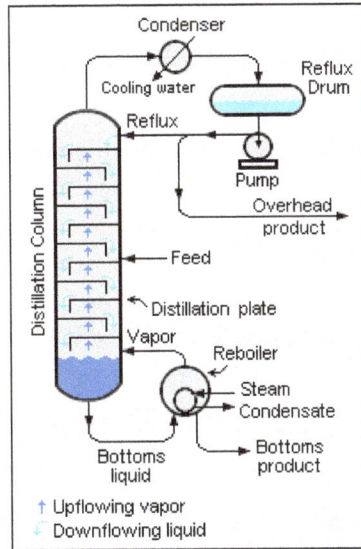

Diagram of a typical industrial distillation tower

Industrial towers use reflux to achieve a more complete separation of products. Reflux refers to the portion of the condensed overhead liquid product from a distillation or fractionation tower that is returned to the upper part of the tower as shown in the schematic diagram of a typical, large-scale industrial distillation tower. Inside the tower, the downflowing reflux liquid provides cooling and condensation of the upflowing vapors thereby increasing the efficiency of the distillation tower. The more reflux that is provided for a given number of theoretical plates, the better the tower's separation of lower boiling materials from higher boiling materials. Alternatively, the more reflux that is provided for a given desired separation, the fewer the number of theoretical plates required. Chemical engineers must choose what combination of reflux rate and number of plates is both economically and physically feasible for the products purified in the distillation column.

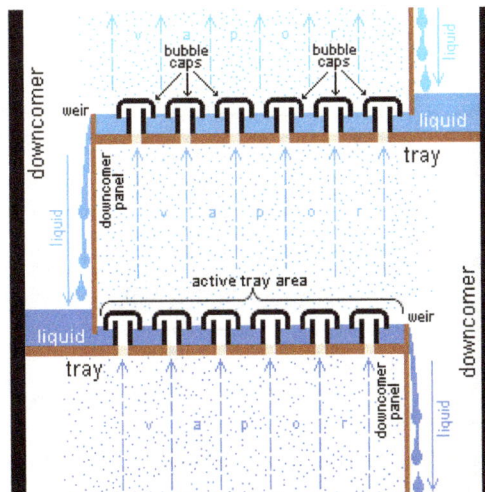

Section of an industrial distillation tower showing detail of trays with bubble caps

Such industrial fractionating towers are also used in cryogenic air separation, producing liquid oxygen, liquid nitrogen, and high purity argon. Distillation of chlorosilanes also enables the production of high-purity silicon for use as a semiconductor.

Design and operation of a distillation tower depends on the feed and desired products. Given a simple, binary component feed, analytical methods such as the McCabe–Thiele method or the Fenske equation can be used. For a multi-component feed, simulation models are used both for design and operation. Moreover, the efficiencies of the vapor–liquid contact devices (referred to as "plates" or "trays") used in distillation towers are typically lower than that of a theoretical 100% efficient equilibrium stage. Hence, a distillation tower needs more trays than the number of theoretical vapor–liquid equilibrium stages. A variety of models have been postulated to estimate tray efficiencies.

In modern industrial uses, a packing material is used in the column instead of trays when low pressure drops across the column are required. Other factors that favor packing are: vacuum systems, smaller diameter columns, corrosive systems, systems prone to foaming, systems requiring low liquid holdup, and batch distillation. Conversely, factors that favor plate columns are: presence of solids in feed, high liquid rates, large column diameters, complex columns, columns with wide feed composition variation, columns with a chemical reaction, absorption columns, columns limited by foundation weight tolerance, low liquid rate, large turn-down ratio and those processes subject to process surges.

Large-scale, industrial vacuum distillation column

This packing material can either be random dumped packing (1–3" wide) such as Raschig rings or structured sheet metal. Liquids tend to wet the surface of the packing and the vapors pass across this wetted surface, where mass transfer takes place. Unlike conventional tray distillation in which every tray represents a separate point of vapor–liquid equilibrium, the vapor–liquid equilibrium curve in a packed column is continuous. However, when modeling packed columns, it is useful to compute a number of "theoretical stages" to denote the separation efficiency of the packed column with respect to more traditional trays. Differently shaped packings have different surface areas and void space between packings. Both of these factors affect packing performance.

Another factor in addition to the packing shape and surface area that affects the performance of random or structured packing is the liquid and vapor distribution entering the packed bed. The number of theoretical stages required to make a given separation is calculated using a specific vapor to liquid ratio. If the liquid and vapor are not evenly distributed across the superficial tower area as it enters the packed bed, the liquid to vapor ratio will not be correct in the packed bed and the required separation will not be achieved. The packing will appear to not be working properly. The height equivalent to a theoretical plate (HETP) will be greater than expected. The problem is not the packing itself but the mal-distribution of the fluids entering the packed bed. Liquid mal-distribution is more frequently the problem than vapor. The design of the liquid distributors used to introduce the feed and reflux to a packed bed is critical to making the packing perform to it maximum efficiency. Methods of evaluating the effectiveness of a liquid distributor to evenly distribute the liquid entering a packed bed can be found in references. Considerable work as been done on this topic by Fractionation Research, Inc. (commonly known as FRI).

Multi-effect Distillation

The goal of multi-effect distillation is to increase the energy efficiency of the process, for use in desalination, or in some cases one stage in the production of ultrapure water. The number of effects is inversely proportional to the kW·h/m³ of water recovered figure, and refers to the volume of water recovered per unit of energy compared with single-effect distillation. One effect is roughly 636 kW·h/m³.

- Multi-stage flash distillation can achieve more than 20 effects with thermal energy input, as mentioned in the article.

- Vapor compression evaporation commercial large-scale units can achieve around 72 effects with electrical energy input, according to manufacturers.

There are many other types of multi-effect distillation processes, including one referred to as simply multi-effect distillation (MED), in which multiple chambers, with intervening heat exchangers, are employed.

Distillation in Food Processing

Distilled Beverages

Carbohydrate-containing plant materials are allowed to ferment, producing a dilute solution of ethanol in the process. Spirits such as whiskey and rum are prepared by distilling these dilute solutions of ethanol. Components other than ethanol, including water, esters, and other alcohols, are collected in the condensate, which account for the flavor of the beverage. Some of these beverages are then stored in barrels or other containers to acquire more flavor compounds and characteristic flavors.

Distillation refers to the physical separation of a mixture into two or more fractions that have different boiling points. The general objective is to separate substances having different vapor pressures at any given temperature. If a liquid mixture of two volatile materials is heated, the vapor that comes out with a higher concentration of the lower boiling material than the liquid from which it was evolved. Conversely, if a warm vapor is cooled, the higher boiling material has a tendency to condense in a greater proportion than the lower boiling material. The early distillers of alcohol for beverages applied these fundamental ideas.

A distillation column consists of a series of plates (or trays). In normal operation, there is a certain amount of liquid on each plate, and some arrangement is made for ascending vapors to pass through the liquid and make contact with it. The descending liquid flows down from the plate above through a downcomer, across the next plate, and then over a weir and into another downcomer to the next lower plate as shown in figure. Earlier on, bubble caps were used on the trays for the purpose of vapor and liquid contacting. Recent developments include the use of sieve tarys, valve trays, perforated or ballast trays.

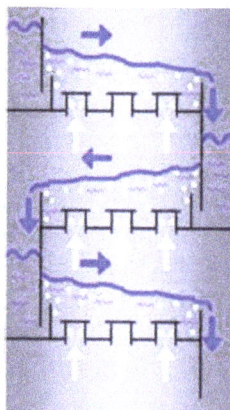

Section of Distillation Column

Distillation columns can be as high as 200ft. Diameters as large as 44 ft have been used; operates at a pressure ranging from 15mm to 500psia. As indicated in figure, the overhead vapor V_1, upon leaving the top plate enters the condenser where it is either partially or totally condensed. The liquid formed is collected in an accumulator from which the liquid stream L_0 (reflux) and the top product stream D (distillate) are withdrawn. When the overhead vapor, V_1 is totally condensed to the liquid state to produce

L_o and D is withdrawn as a liquid, the condenser is called total condenser. If V_1 is partially condensed to the liquid state to produce L_o and D is withdrawn as a vapor, the condenser is called partial condenser. The amount of reflux is generally expressed in terms of reflux ratio, L_o/D. The liquid that leaves the bottom plate of the column enters the reboiler, where it is partially vaporized. The vapor produced is allowed to flow back up through the column, and the liquid is withdrawn from the reboiler known as bottom product, B.

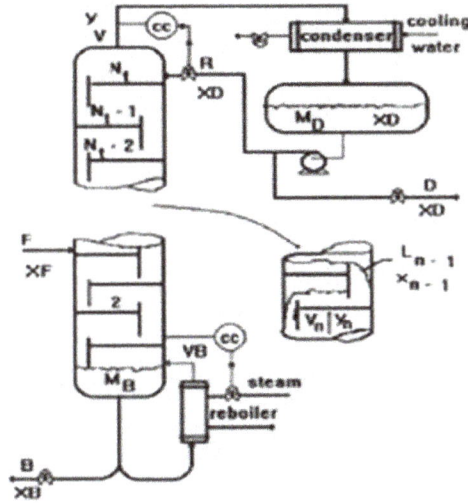

Continuous Distillation Column

Fundamental Principles Involved in Distillation

In order to compute top product, D and bottom product, B ; it is necessary to obtain a solution of equilibrium relationships, component material balances, total material balances and energy balances respectively.

Physical Equilibrium

A two phase multicomponent mixture is said to be in equilibrium if

- The temperature T^v of the vapor phase is equal to T^l of the liquid phase.

- The total pressure P^v throughout the vapor phase is equal to the total pressure P^l throughout the liquid phase.

- The tendency of each component to escape from the liquid phase to the vapor phase is exactly equal to its tendency to escape from the vapor phase to the liquid phase.

In the following analysis it is assumed that $T^v = T^l = T$, $P^v = P^l = P$ and the escaping tendencies are equal. A special case of the third condition for equilibrium is represented by Raoult's law.

$$Py_i = P_i x_i \qquad (1)$$

Where x_i and y_i are liquid and vapor mole fractions of component i at temperature T of the system.

The separation of a binary mixture is represented by a 2D space. The graphical method by McCabe and Thiele for solution of problems involving binary mixtures is shown in subsequent section. This method makes use of an equilibrium curve that may be obtained from the "boiling point diagram"(BPD).

Construction and interpretation of BPD for binary mixtures

For a binary system having components A and B; the equilibrium relationships are given as follows: $Py_A = P_A x_A$, $Py_B = P_A x_B$, $y_A + y_B = 1$, $x_A + x_B = 1$. $\qquad (2)$

Following Gibbs phase rule, the degree of freedom for the above set of equations (2) is 2 and thus two variables must be fixed in order to get the solution of the equation. For construction of BPD, the total pressure P must be fixed and a solution is obtained for each of several temperatures lying between the temperatures at which the respective vapor pressures P_A and P_B are equal to the total pressure P. The solution of eqation for x_A in terms of P_A and P_B and P is effected as follows

$$P = P_A x_A + P_B x_B \qquad (3)$$

$$Or, x_A = (P - P_B)/(P_A - P_B) \qquad (4)$$

From the definition of a mole fraction, equation has a meaningful solution at a given P for any T lying between the boiling point temperatures T_A and T_B of pure A and pure B. Using the first expression of equation y_A can also be calculated. By plotting T vs x_A and T vs y_A we get the lower and upper curves respectively, figure are typical of those obtained when when component A is more volatile than B. If for the close interval $T_A \leq T \leq T_B, P_A > P_B$, the parallel lines such as CE that join the equilibrium pairs (x,y), computed at a given values of T and P by the earlier quations is called Tie lines.

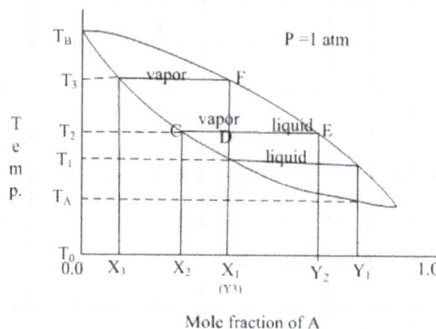

Mole fraction of A

The boiling point Diagram

Determination of bubble point and dew point temperatures of multi- component mixtures (BPT, DPT)

The state of equilibrium for a two phase system is described by the following equations in which any number of components 'c' are distributed between the two phases

$$y_i = K_i x_i, \quad \sum_{i=1}^{c} y_i = 1, \quad \sum_{i=1}^{c} x_i = 1 \quad where(1 \le i \le c) \tag{5}$$

where K_i is ratio of the fugacities of ith component at liquid to vapor phases respectively for an ideal solution. From this equation it is evident that number of equations is 'c+2' and number of variables is '2c+2'. Thus, to obtain a solution to these equations, 'c' variables must be fixed.

When the first expression of Equation (5) is summed over for all components, we get

$$1 = \sum_{i=1}^{c} K_i x_i \tag{6}$$

where K_i is an implicit function of temperature. The solution of the last equation is achieved by Newton's method. In this method, the Equation (6) is restated as shown

$$f(T) = \sum_{i=1}^{c} K_i x_i - 1$$

The value of T at which f(T) =0, is the bubble point. In this iterative scheme we need to find the first derivative of f(T) and further follow the iterative form

$T_{n+1} = T_n - f(T_n)/f'(T_n)$. where T_n is the assumed value for the first stage of the iteration and T_{n+1} is the modified value after the iteration. The process goes on as long as the values of T are not within a convergence criteria.

When the y_i's and P are fixed rather than x_i's and P, the solution temperature of the Eq (5) will give the DBT. Here, just like previous case the governing equation is

$$F(T) = \sum_{i=1}^{c} y_i / K_i - 1.$$

K_b's method of calculating DBT AND BPT

Robinson and Gilliland pointed out that if the relative values of the K_i's are independent of temperature, the Eq (5) may be rearranged so that the iterative methods used in previous case may be avoided. Ratio K_i/K_b is called the relative volatility, i, of component 'i' with respect to component 'b'; where the K values are calculated at same temperature and pressure.

For calculation of BPT, Equationis (5) rewritten as

$$y_i = (K_i / K_b)K_b x_i = \alpha_1 K_b x_i \tag{7}$$

Taking summation on both sides, we get on rearranging

$$K_b = 1/(\sum_{i=1}^{c} K_i \alpha_i) \tag{8}$$

Since α_i's are independent of temperature, they maybe computed by K_i & K_b evaluated at any arbitrary T and at specified pressure. After K_b has been calculated using eq(8), the BPT can be found from the known relation between K_b and T. The same procedure is obtained for calculation of DPT.

Separation of Multicomponent Mixtures by use of a Single Equilibrium Stage

Each of the separation processes considered here are special cases of the general separation problem in which a multi-component mixture is to be separated into two or more parts through the use of any number of equilibrium stages.

Flash Calculations

The boiling point diagram is useful for visualizing the necessary conditions required for a flash to occur. Suppose the feed to be flashed has the composition x_i, and further suppose that this liquid must be at the temperature T_0 and pressure, P = 1atm, is to be flashed by raising the temperature to the specified flash temperature $T_F = T_2$ at specified flash pressure, P = 1atm. It is observed that the BPT of the feed T_{BPT} = at P = 1atm is T_1. The DPT, T_{DPT} of the feed at the pressure 1 atm is T_3. Then a necessary condition for a flash to occur at the specified pressure is that ; $T_{BPT} < T_F < T_{DPT}$

In practice, the flash process is generally carried out by reducing the pressure on the feed stream rather than heating the feed at constant pressure. To determine whether or not the feed will flash at a given T_F & P, the following two conditions must be satisfied:

1. $f(T_F) > 0, F(T_F) > 0$ \hfill (9)

2. $f(T_F) = \sum_{i=1}^{c} K_{Fi} X_i - 1, \quad F(T_F) = \sum_{i=1}^{c} X_i / K_{Fi} - 1$ \hfill (10)

There are two types of flash calculations : isothermal flash and adiabatic flash.

Isothermal Flash

In isothermal flash the following specifications are made; T_F, P, $\{X_i\}$ and F. We have to find the unknowns V_F, L_F, $\{y_{Fi}\}$ and $\{x_{Fi}\}$. The following equations are needed:

Equilibrium relations :

$$y_{Fi} = K_{Fi}x_{Fi},, \quad \sum_{i=1}^{c} y_{Fi} = 1, \quad \sum_{i=1}^{c} x_{Fi} = 1 \quad where(1 \le i \le c) \tag{11}$$

Material balances: $Fx_i = V_F y_{Fi} + L_F x_{Fi}$.

Eq(10) is seen to represent 2c+2 equations in 2c+2 unknowns. This system of non-linear equations is readily reduced to one equation in one unknown (say V_F) in the following manner. First observe that the total material balance expression may be obtained by summing each member of the last expression of Eq (9) over all components to give,

$$F \sum_{i=1}^{c} X_1 = V_F \sum_{i=1}^{c} y_{Fi} + L_F \sum_{i=1}^{c} x_{Fi} \quad or \quad F = V_F + L_F \tag{12}$$

Elimination of y_{Fi}'s from the last expression of Eq (9) by use of the first expression, followed by rearrangement gives :

$$x_{Fi} = X_i / (L_F / F + V_F K_{FV} / F) \tag{13}$$

Elimination of L_F from Eq (11) by eq (10) yields

$$x_{Fi} = X_i / (1 - \psi(1 - K_{Fi})) \tag{14}$$

Where $\psi = V_F / F$.

when each member of Eq (12) is summed over all components and the result so obtained is restated in functional notation, one obtains

$$P(\psi) = \sum_{i=1}^{c} X_i / (1 - \psi(1 - K_{Fi})) - 1 \tag{15}$$

$$and \quad P'(\psi) = \sum_{i=1}^{c} X_i (1 - K_{Fi}) / (1 - \psi(1 - K_{Fi}))^2 \tag{16}$$

$$h = -P(\psi) / P'(\psi)$$

$$P^{k+1}(\psi) = P^k(\psi) + h$$

The specification of T_F implies that the feed either posses precisely the correct amount of energy for the flash to account at T_F at specified P or, that, energy is to be added or withdrawn at the flash drum.

Enthalpy balance:

$$FH = V_F H_F + L_F h_F \tag{17}$$

$$H_F = \sum_{i=1}^{c} H_{Fi} y_{Fi}, \quad h_F = \sum_{i=1}^{c} h_{Fi} X_{Fi} \tag{18}$$

Adiabatic Flash

The term adiabatic flash is used to describe the problem wherein the following specifications are made; P,Q = 0 (no heat is added at flash drum), H, {X$_i$}, F. In this case there are 2c+3 unknowns [T$_F$, V$_F$,L$_F$,{y$_{Fi}$},{x$_{Fi}$}] and 2c+3 independent equations.

Equilibrium relations, material balances and energy balances are same as in previous case. To solve this problem using N-R method, it is essential that number of independent functions to be equal to number of independent variables.

$$f_1 = K_{F1}x_{F1} - y_{F1}$$

$$f_2 = K_{F2}x_{F2} - y_{F2}$$

$$\dotsb$$

$$f_c = K_{Fc}x_{Fc} - y_{Fc}$$

$$f_{c+1} = \sum_{i=1}^{c} y_{Fi} - 1$$
$$f_{c+2} = \sum_{i=1}^{c} x_{Fi} - 1$$

$$f_{c+3} = V_F y_{F1} + L_F x_{F1} - FX_1$$

$$f_{c+4} = V_F y_{F2} + L_F x_{F2} - FX_2$$

$$\dotsb$$

$$f_{2c+2} = V_F y_{Fc} + L_F x_{Fc} - FX_c$$

$$f_{2c+3} = V_F H_F + L_F h_F - FH$$

The application of the N-R method to this set of equations may be represented by the following matrix equation : $J\Delta x = -f$

The jacobian matrix, J and the column vectors Δx are defined as follows:

$$J = \begin{bmatrix} \dfrac{\partial f1}{\partial yf1} & \cdots & \dfrac{\partial f1}{\partial yfc} & \dfrac{\partial f1}{\partial xf1} & \cdots & \dfrac{\partial f1}{\partial xfc} & \dfrac{\partial f1}{\partial VF} & \dfrac{\partial f1}{\partial LF} & \dfrac{\partial f1}{\partial TF} \\ \vdots & \ddots & & & & \vdots & & & \\ \dfrac{\partial f2c+3}{\partial yf1} & \cdots & \dfrac{\partial f2c+3}{\partial yfc} & \dfrac{\partial f2c+3}{\partial xf1} & \cdots & \dfrac{\partial f2c+3}{\partial xfc} & \dfrac{\partial f2c+3}{\partial VF} & \dfrac{\partial f2c+3}{\partial LF} & \dfrac{\partial f2c+3}{\partial TF} \end{bmatrix}$$

$$\Delta x = [\Delta y_{F1}......\Delta y_{Fc} \quad \Delta x_{F1}......\Delta x_{Fc} \quad \Delta V_F \quad \Delta L_F \quad \Delta T_F]^T$$

$$f = [f_1 f_2......f_c \quad f_{c+1}\cdots ...f_{2c} \quad f_{2c+1}....f_{2c+3}]^T$$

where each element of Δx is equal to the newly predicted values of the variable minus the assumed value; for example, $\Delta y_{F1} = y_{F1,n+1} - y_{F1,n}.$ To initiate the calculations, a complete set of values for the variables must be assumed; say $(y_{F1,n}.....y_{Fc,n} \quad X_{F1,n}.....X_{Fc,n} \quad V_{F,n} \quad L_{F,n} \quad T_{F,n}).$

Alternative Process for Adiabatic Flash

When the pressure of a liquid stream of known composition, flowrate, and temperature is reduced adiabatically across a value an adiabatic flash calculation is made to determine the resulting temperature, compositions and flowrates of equilibrium liquid and vapor streams for a specified pressure downstream of the value. For an adiabatic flash, the isothermal flash calculation procedure can be applied in the iterative way. A guess is made of the flash temperature T_v. Then φ, V, x, y and L are determined as for an isothermal flash. The guessed value of T_v (equal to T_l) is next checked by an energy balance with Q=0 to give

$$F(T_v) = (\varphi_{hv} + (1-\varphi)h_1 - h_f)/1000 = 0 \tag{19}$$

Where the division by 1000 is done in order to convert the terms in the order of 1. If the computed value of F (T_v) is not zero, the entire procedure is repeated for two or more T_v. The procedure is tedious as it involves inner loop iteration on φ and outer loop iteration of T_v. outer loop iteration is successful when equation

$$f(\varphi) = 0 = \sum_{i=1}^{c} (z_i)(1-K_i)/(1+\varphi(K_i-1))$$

$$\text{Where } \varphi = \frac{V}{F}, K_i \text{ is a function of } \{T_v, P_v\} \text{ and } V = \varphi F$$

For the closed boiling mixers, the above algorithm dose not work well due to extreme sensitivity T_s in the inner loop. The above problem can be remove by first assume T_v and iterated for external loop and inside loop be iterated for φ.

$$F(T_v) = \sum_{i=1}^{c} (z_i)(1-K_i)/(1+\varphi(K_i-1)) \tag{20}$$

Then compute x & y from the following equations:-

$$X_i = z_i / (1 + \varphi(K_v - 1))$$
$$Y_i = z_i k_i / (1 + \varphi(K_v - 1)) = x_i k_i$$

Eq (1) is solved directly for φ, from which

$$\varphi = (h_f - h_1) / (h_v - h_1) \tag{21}$$

If φ from Eq(21) is not equal to assumed φ then solve Eq(20), the new value of φ is to repeat the outer loop starting with Eq(20).

Rachford-Rice Procedure for isothermal flash calculations when K-values are independent of composition follows the algorithm as stated below

Step 1. Specify variables $F, T_F, P_F, X_1, X_2, ..., X_c, T_v, P_v$

Step 2. $T_L = T_V, P_L = P_V$

Step 3. Solve $\{\varphi\} = \sum_{i=1}^{c} (X_i)(1 - K_i) / (1 + \varphi(K_i - 1))$ Where $\varphi = \dfrac{V}{F}$ K_i is a function of $\{T_V, P_V\}$ and $V = \varphi F$

Step 4. $X_i = z_i / (1 + \varphi(K_v - 1)), Y_i = z_i k_i / (1 + \varphi(K_v - 1)) = x_i k_i$

Step 5. $L = F - V$

Step 6. $Q = h_V V + h_L L - h_F F$ is set to 0

Using step 5, and further simplifying, we get $\varphi h_v + (1 - \varphi) h_L - h_F = 0$

We may use softwares like Chemcad design II, PRO/II, ASPEN PLUS to solve these equations.

Continuous Distillation

Continuous distillation, a form of distillation, is an ongoing separation in which a mixture is continuously (without interruption) fed into the process and separated fractions are removed continuously as output streams. Distillation is the separation or partial separation of a liquid feed mixture into components or fractions by selective boiling (or evaporation) and condensation. The process produces at least two output fractions. These fractions include at least one *volatile* distillate fraction, which has boiled and been separately captured as a vapor condensed to a liquid, and practically always a *bottoms* (or *residuum*) fraction, which is the least volatile residue that has not been separately captured as a condensed vapor.

An alternative to continuous distillation is batch distillation, where the mixture is added to the unit at the start of the distillation, distillate fractions are taken out sequentially in time (one after another) during the distillation, and the remaining bottoms fraction is removed at the end. Because each of the distillate fractions are taken out at different times, only one distillate exit point (location) is needed for a batch distillation and the distillate can just be switched to a different receiver, a fraction-collecting container. Batch distillation is often used when smaller quantities are distilled. In a continuous distillation, each of the fraction streams is taken simultaneously throughout operation; therefore, a separate exit point is needed for each fraction. In practice when there are multiple distillate fractions, the distillate exit points are located at different heights on a fractionating column. The bottoms fraction can be taken from the bottom of the distillation column or unit, but is often taken from a reboiler connected to the bottom of the column.

Each fraction may contain one or more components (types of chemical compounds). When distilling crude oil or a similar feedstock, each fraction contains many components of similar volatility and other properties. Although it is possible to run a small-scale or laboratory continuous distillation, most often continuous distillation is used in a large-scale industrial process.

Industrial Application

Distillation is one of the unit operations of chemical engineering. Continuous distillation is used widely in the chemical process industries where large quantities of liquids have to be distilled. Such industries are the natural gas processing, petrochemical production, coal tar processing, liquor production, liquified air separation, hydrocarbon solvents production and similar industries, but it finds its widest application in petroleum refineries. In such refineries, the crude oil feedstock is a very complex multicomponent mixture that must be separated and yields of pure chemical compounds are not expected, only groups of compounds within a relatively small range of boiling points, which are called *fractions*. These *fractions* are the origin of the term *fractional distillation* or *fractionation*. It is often not worthwhile separating the components in these fractions any further based on product requirements and economics.

Industrial distillation is typically performed in large, vertical cylindrical columns known as "distillation towers" or "distillation columns" with diameters ranging from about 65 centimeters to 11 meters and heights ranging from about 6 meters to 60 meters or more.

Principle

The principle for continuous distillation is the same as for normal distillation: when a liquid mixture is heated so that it boils, the composition of the vapor above the liquid differs from the liquid composition. If this vapor is then separated and condensed into a liquid, it becomes richer in the lower boiling point component(s) of the original mixture.

This is what happens in a continuous distillation column. A mixture is heated up, and routed into the distillation column. On entering the column, the feed starts flowing down but part of it, the component(s) with lower boiling point(s), vaporizes and rises. However, as it rises, it cools and while part of it continues up as vapor, some of it (enriched in the less volatile component) begins to descend again.

Image 3 depicts a simple continuous fractional distillation tower for separating a feed stream into two fractions, an overhead distillate product and a bottoms product. The "lightest" products (those with the lowest boiling point or highest volatility) exit from the top of the columns and the "heaviest" products (the bottoms, those with the highest boiling point) exit from the bottom of the column. The overhead stream may be cooled and condensed using a water-cooled or air-cooled condenser. The bottoms reboiler may be a steam-heated or hot oil-heated heat exchanger, or even a gas or oil-fired furnace.

In a continuous distillation, the system is kept in a steady state or approximate steady state. Steady state means that quantities related to the process do not change as time passes during operation. Such constant quantities include feed input rate, output stream rates, heating and cooling rates, reflux ratio, and temperatures, pressures, and compositions at every point (location). Unless the process is disturbed due to changes in feed, heating, ambient temperature, or condensing, steady state is normally maintained. This is also the main attraction of continuous distillation, apart from the minimum amount of (easily instrumentable) surveillance; if the feed rate and feed composition are kept constant, product rate and *quality* are also constant. Even when a variation in conditions occurs, modern process control methods are commonly able to gradually return the continuous process to another steady state again.

Since a continuous distillation unit is fed constantly with a feed mixture and not filled all at once like a batch distillation, a continuous distillation unit does not need a sizable distillation pot, vessel, or reservoir for a batch fill. Instead, the mixture can be fed directly into the column, where the actual separation occurs. The height of the feed point along the column can vary on the situation and is designed so as to provide optimal results.

A continuous distillation is often a fractional distillation and can be a vacuum distillation or a steam distillation.

Design and Operation

Design and operation of a distillation column depends on the feed and desired products. Given a simple, binary component feed, analytical methods such as the McCabe–Thiele method or the Fenske equation can be used to assist in the design. For a multi-component feed, computerized simulation models are used both for design and subsequently in operation of the column as well. Modeling is also used to optimize already erected columns for the distillation of mixtures other than those the distillation equipment was originally designed for.

When a continuous distillation column is in operation, it has to be closely monitored for changes in feed composition, operating temperature and product composition. Many of these tasks are performed using advanced computer control equipment.

Column Feed

The column can be fed in different ways. If the feed is from a source at a pressure higher than the distillation column pressure, it is simply piped into the column. Otherwise, the feed is pumped or compressed into the column. The feed may be a superheated vapor, a saturated vapor, a partially vaporized liquid-vapor mixture, a saturated liquid (i.e., liquid at its boiling point at the column's pressure), or a sub-cooled liquid. If the feed is a liquid at a much higher pressure than the column pressure and flows through a pressure let-down valve just ahead of the column, it will immediately expand and undergo a partial flash vaporization resulting in a liquid-vapor mixture as it enters the distillation column.

Improving Separation

Although small size units, mostly made of glass, can be used in laboratories, industrial units are large, vertical, steel vessels known as "distillation towers" or "distillation columns". To improve the separation, the tower is normally provided inside with horizontal plates or trays, or the column is packed with a packing material. To provide the heat required for the vaporization involved in distillation and also to compensate for heat loss, heat is most often added to the bottom of the column by a *reboiler*, and the purity of the *top product* can be improved by recycling some of the externally condensed top product liquid as reflux. Depending on their purpose, distillation columns may have liquid outlets at intervals up the length of the column.

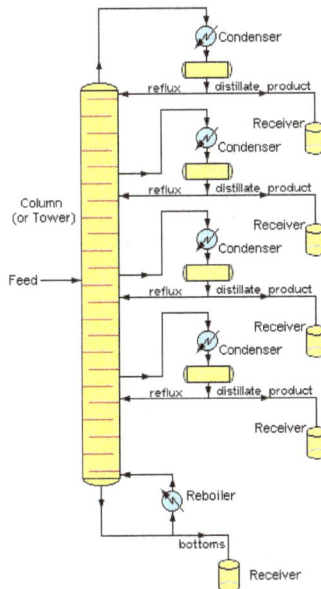

Simplified chemical engineering schematic of Continuous Fractional Distillation tower separating one feed mixture stream into four distillate and one bottoms fractions

Reflux

Large-scale industrial fractionation towers use reflux to achieve more efficient separation of products. Reflux refers to the portion of the condensed overhead liquid product from a distillation tower that is returned to the upper part of the tower as shown in images 3 and 4. Inside the tower, the downflowing reflux liquid provides cooling and partial condensation of the upflowing vapors, thereby increasing the efficacy of the distillation tower. The more reflux that is provided, the better is the tower's separation of the lower boiling from the higher boiling components of the feed. A balance of heating with a reboiler at the bottom of a column and cooling by condensed reflux at the top of the column maintains a temperature gradient (or gradual temperature difference) along the height of the column to provide good conditions for fractionating the feed mixture. Reflux flows at the middle of the tower are called pumparounds.

Changing the reflux (in combination with changes in feed and product withdrawal) can also be used to improve the separation properties of a continuous distillation column while in operation (in contrast to adding plates or trays, or changing the packing, which would, at a minimum, require quite significant downtime).

Plates or Trays

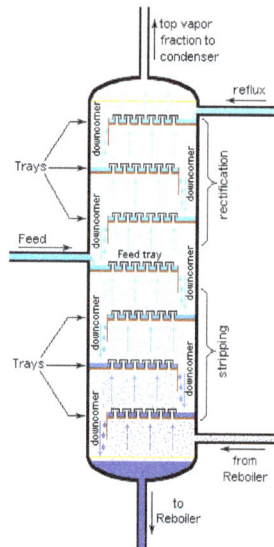

Cross-sectional diagram of a binary fractional distillation tower with bubble-cap trays.

Distillation towers use various vapor and liquid contacting methods to provide the required number of equilibrium stages. Such devices are commonly known as "plates" or "trays". Each of these plates or trays is at a different temperature and pressure. The stage at the tower bottom has the highest pressure and temperature. Progressing upwards in the tower, the pressure and temperature decreases for each succeeding stage. The vapor–liquid equilibrium for each feed component in the tower reacts in its unique way to the different pressure and temperature conditions at each of the stages. That means that each component establishes a different concentration in the vapor and liquid phases at each of the stages, and this results in the separation of the components. Some example trays are depicted in image 5. A more detailed, expanded image of two trays can be seen in the theoretical plate article. The reboiler often acts as an additional equilibrium stage.

If each physical tray or plate were 100% efficient, then the number of physical trays needed for a given separation would equal the number of equilibrium stages or theoretical plates. However, that is very seldom the case. Hence, a distillation column needs more plates than the required number of theoretical vapor–liquid equilibrium stages.

Packing

Another way of improving the separation in a distillation column is to use a packing material instead of trays. These offer the advantage of a lower pressure drop across the column (when compared to plates or trays), beneficial when operating under vacuum. If a distillation tower uses packing instead of trays, the number of necessary theoretical equilibrium stages is first determined and then the packing height equivalent to a theoretical equilibrium stage, known as the *height equivalent to a theoretical plate* (HETP), is also determined. The total packing height required is the number theoretical stages multiplied by the HETP.

This packing material can either be random dumped packing such as Raschig rings or structured sheet metal. Liquids tend to wet the surface of the packing and the vapors pass across this wetted surface, where mass transfer takes place. Unlike conventional tray distillation in which every tray represents a separate point of vapor–liquid equilibrium, the vapor–liquid equilibrium curve in a packed column is continuous. However, when modeling packed columns it is useful to compute a number of theoretical plates to denote the separation efficiency of the packed column with respect to more traditional trays. Differently shaped packings have different surface areas and void space between packings. Both of these factors affect packing performance.

Another factor in addition to the packing shape and surface area that affects the performance of random or structured packing is liquid and vapor distribution entering the packed bed. The number of theoretical stages required to make a given separation is calculated using a specific vapor to liquid ratio. If the liquid and vapor are not evenly distributed across the superficial tower area as it enters the packed bed, the liquid to vapor ratio will not be correct in the packed bed and the required separation will not be achieved. The packing will appear to not be working properly. The *height equivalent to a theoretical plate* (HETP) will be greater than expected. The problem is not the packing itself but the mal-distribution of the fluids entering the packed bed. Liquid mal-distribution is more frequently the problem than vapor. The design of the liquid distributors used to introduce the feed and reflux to a packed bed is critical to making the packing perform at maximum efficiency. Methods of evaluating the effectiveness of a liquid distributor can be found in references.

Overhead System Arrangements

Overhead stream that is totally condensed into a liquid product using water or air-cooling. However, in many cases, the tower overhead is not easily condensed totally and the reflux drum must include a vent gas outlet stream. In yet other cases, the overhead stream may also contain water vapor because either the feed stream contains some water or some steam is injected into the distillation tower (which is the case in the crude oil distillation towers in oil refineries). In those cases, if the distillate product is insoluble in water, the reflux drum may contain a condensed liquid distillate phase, a condensed water phase and a non-condensible gas phase, which makes it necessary that the reflux drum also have a water outlet stream.

Examples

Continuous Distillation of Crude Oil

Petroleum crude oils contain hundreds of different hydrocarbon compounds: paraffins, naphthenes and aromatics as well as organic sulfur compounds, organic nitrogen compounds and some oxygen containing hydrocarbons such as phenols. Although crude oils generally do not contain olefins, they are formed in many of the processes used in a petroleum refinery.

The crude oil fractionator does not produce products having a single boiling point; rather, it produces fractions having boiling ranges. For example, the crude oil fractionator produces an overhead fraction called "naphtha" which becomes a gasoline component after it is further processed through a catalytic hydrodesulfurizer to remove sulfur and a catalytic reformer to reform its hydrocarbon molecules into more complex molecules with a higher octane rating value.

The naphtha cut, as that fraction is called, contains many different hydrocarbon compounds. Therefore, it has an initial boiling point of about 35 °C and a final boiling point of about 200 °C. Each cut produced in the fractionating columns has a different boiling range. At some distance below the overhead, the next cut is withdrawn from the side of the column and it is usually the jet fuel cut, also known as a kerosene cut. The boiling range of that cut is from an initial boiling point of about 150 °C to a final boiling point of about 270 °C, and it also contains many different hydrocarbons. The next cut further down the tower is the diesel oil cut with a boiling range from about 180 °C to about 315 °C. The boiling ranges between any cut and the next cut overlap because the distillation separations are not perfectly sharp. After these come the heavy fuel oil cuts and finally the bottoms product, with very wide boiling ranges. All these cuts are processed further in subsequent refining processes.

The numerical simulation of multi-component separation process requires the following points:-

- ❖ Reliable predictive method for computing phase equilibrium, more particularly for calculating equilibrium constant & enthalpies.

- ❖ A realistic mathematical model with good flexibility & generalization properties.

- ❖ A reliable and efficient numerical method for computing the solution.

- ❖ The capability of using the method for design purposes.

Steady state modeling:-

The mathematical relations describing the counter current separation process like distillation consists of four sets of basic equations, called mesh equations:

- ❖ Mass balance equation(M)

- ❖ Equilibrium relations(E)

- ❖ Sum(or conservation) equation(S)

- ❖ Enthalpy balance relations(H)

Presentation and arrangement depend on the chosen iterative variable and on the balance formulation; so various numerical methods have been proposed. An analysis of their advantages and disadvantages was reviewed by Wang et.al.(1981). The method can be classified into two main categories:-

- ❖ Partitioning and coupling methods.

- ❖ Global and simultaneous procedures.

In each class the algorithms can differ from one another according to the following criteria:-

1. The iterative variable chosen and mathematical model used.

2. The way of arranging the equations and the procedure implemented for their solution.

3. The convergence method and the convergence promoters.

Partitioning and coupling methods: -The mesh equations are grouped either by stage or by type. Initials estimates for some variables are required for starting the calculations. For a given set of variables the groups of equations are solved in a predetermined order, the other variables being kept constant. These last variables are updated according to the previous result by using direct substitution method. This procedure is continued until a convergence is achieved. Within this category the different approaches can be used.

1. Stage to stage methods: - In this class of methods, the procedure of Lewis and Matheson, Thiele and Geddes, and the Θ method of Holland are used. The mesh equations are grouped stage to stage and solved from bottom to top.

2. **Partitioned by type of equation: -** in this approach temperature and flow rates must be initialised. The mass balance equations are combined with the equilibrium equations to form the first set of equations. These equations are linearised by keeping the constant flow rates and equilibrium constants, and then they are solved component by component. By using the new values of compositions, the sum and enthalpy balance equations can be solved either simultaneously or separately in order to obtain the new temperature & flow rates profiles.

Global and simultaneous procedures: -

In this class of methods, the mesh equations are linearized and simultaneously solved by means of Newtons-Raphson Method. A Jacobian matrix containing a large number of partial derivatives (numerically or analytically) must be generated and the equations are solved with the initial set of variable.

Mathematical model: -

Consider the column shown in following figure, involves N plates numbered from top to bottom, including a condenser (total or partial) numbered 1(one) and a reboiler numbered N. The liquid (molar flow L_j) and vapour (molar flow V_j) coming from the stage j from a liquid – vapour equilibrium. On each stage j, a feed stream F_j and liquid (SL_j) and vapour (SV_j) side stream are planned. Heat supplies or losses Q_j on each plate are taken into account, so non- adiabatic column can be studied.

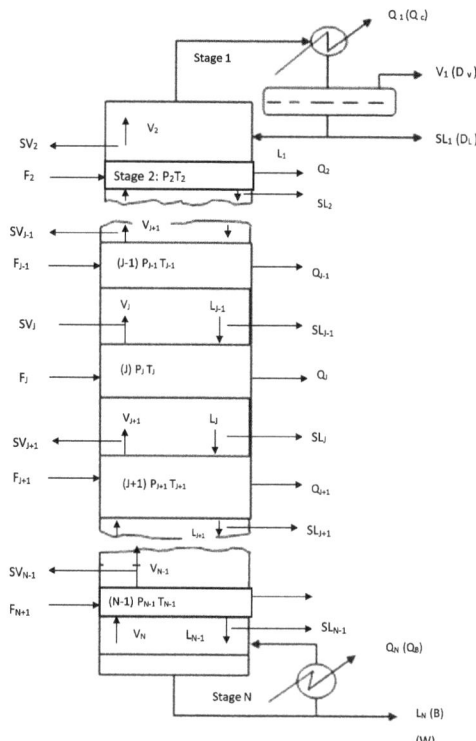

Continuous Distillation Column

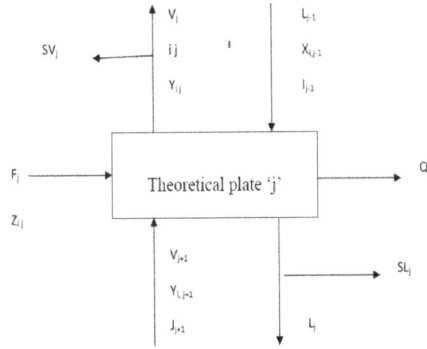

Theoretical Plate of a Continuous Column

When the configuration of the column is fixed, the steady state mesh equations can be expressed as follows for a mixture involving C components:

❖ Mass balance equation(M) (j=1,N)

a) Overall mass balance:

$$L_{j-1} + V_{j+1} + F_j - L_j - SL_j - V_j - SV_j = 0$$

b) Partial mass balance(i=1,C)

$$L_{j-1}.X_{i,j-1} - (L_j + SL_j).X_{ij} - (V_j + SV_j).Y_{ij} + V_{j+1}.Y_{i,j+1} + F_j.Z_{i,j} = 0$$

❖ Equilibrium equations(E) (i=1,C; j=1,N)

$$Y_{ij} - K_{ij}.X_{ij} = 0$$

This equation is based on the assumption that each plate is a theoretical plate, that is to say that liquid vapour equilibrium exists on each plate between the two phases coming from it.

❖ Sum equations(S)

$$\sum_{i=1}^{c}(X_{ij} - Y_{ij}) = 0$$

❖ Enthalpy balance equations (H) (j=1, N)

$$L_{j-1}.H_{j-1} + V_{j+1}.H_{j+1} + F_j.H_{Fj} - (L_j + SL_j).H_j - (V_j + SV_j).H_{vj} - Q_j = 0$$

This enthalpy balance is based on the assumption of an ideal heat transfer, i.e. the vapour and liquid stream coming from each plate have the same temperature. In addition to these equations, and correlations are required to predict the equilibrium constants and the enthalpies. These correlations are often complex equations describing relations between thermodynamic properties and state variable.

Degree of freedom:

In order to define the problem, all the feed flows F_j feed composition Z_{ij} (i=1, C), feed temperatures and pressure are fixed. All the liquid and vapour side stream flows SL_j, SV_j, as well as the amounts of heat supplied Q ($2 \leq j \leq N - 1$) and the pressure P_j are also fixed.

In the case of a total condenser $V_1 = 0$, the heat Q_N supplied to the reboiler is fixed and the reflux ratio, $R = L_1/SL_1$, is also fixed when the side stream flow rate is given. In the case of a partial condenser Q_N and the heat supplied to the condenser Q_1 are fixed. So the unknowns of the problem are liquid and vapour compositions X_{ij}, Y_{ij}, the temperature T_j on each plate, the liquid(L_1 to L_N) and vapour(V_1 to V_N) molar flow i.e t_o.

Mathematical Modeling of Chemical Reactors

Chemical reactor calculations are based on the elementary conservation laws of matter and energy. Information is required on flow phenomenon, rates at which the reactions precede heats of transformation, and description of chemical equilibrium.

Quantitative aspects of chemical Reactions:

Reactions may occur as soon as the reaction mixture is achieved to appropriate conditions of temperature and pressure with reactants well mixed in a molecular scale. Some of them require the use of catalyst which takes part on elementary steps. The catalyst may be under the same state as the reactants (homogeneous catalysts) or in a different state, usually the solid state (heterogeneous catalysts).

Classification of gross reaction types and reaction rates:

From a gross kinetic point of view, chemical reactions are mainly considered under overall formulations which do not refer to elementary processes. There are two classes of reactions

- the single reactions

- the multiple reactions

The single reactions comprise:

1. Uni-directional reactions

2. Reversible reactions

$$v_A^A + V_B^B \rightarrow V_P^P + v_Q^Q$$

$$v_A^A + V_B^B \rightleftharpoons V_P^P + v_Q^Q$$

3. Autocatalytic reactions

$$v_A^A \rightarrow V_P^P$$

Where the presence of the product P is necessary to provide the conservation of the reactant A.

The multiple reactions consist a wide variety of possible kinds of reactions, among which three basic types can be as follows:

- Parallel reactions

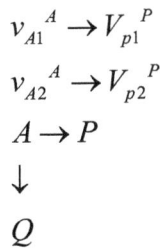

$$v_{A1}{}^A \rightarrow V_{p1}{}^P$$
$$v_{A2}{}^A \rightarrow V_{p2}{}^P$$
$$A \rightarrow P$$
$$\downarrow$$
$$Q$$

Consecutive reactions

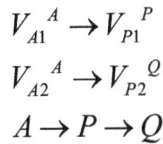

$$V_{A1}{}^A \rightarrow V_{P1}{}^P$$
$$V_{A2}{}^A \rightarrow V_{P2}{}^Q$$
$$A \rightarrow P \rightarrow Q$$

Complex reactions:

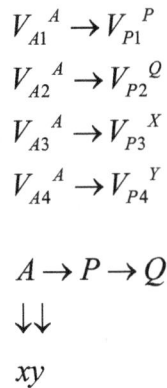

$$V_{A1}{}^A \rightarrow V_{P1}{}^P$$
$$V_{A2}{}^A \rightarrow V_{P2}{}^Q$$
$$V_{A3}{}^A \rightarrow V_{P3}{}^X$$
$$V_{A4}{}^A \rightarrow V_{P4}{}^Y$$

$$A \rightarrow P \rightarrow Q$$
$$\downarrow\downarrow$$
$$xy$$

The reaction formulation can be written under the form of a reaction balance. Thus for the reaction I this form gives:

$$\sum_j v_{ij} A_j = 0$$

Where A_j is the component j which corresponds to either a product or a reactant, and

V_{ij} is the stoichiometry coefficient of the component A_j reaction i. in this description V_{ij} is positive for a product and negative for a reactant.

Single reactions are described by a single reaction balance; multiple reactions are described by a system of reaction balances, the number of which is that of all the reactions involved.

in a single reaction I, the production rate of a given component is A_j defined by:

$$r_{ij} = \frac{1}{V} \frac{\partial N_j}{\partial t}$$

Where, V = volume of reaction mixture

N_j = mole number of the component A_j

From the reaction stoichiometry, it can readily be established that $r_{ij} / V_{ij} /$ is the same for all the components. This rate is called the kinetic rate of the reaction i:

$$r_i = \frac{r_i}{V_{ij}}$$

It can be noted that r_i is positive.

In the case of multiple reactions, the overall production rate of the component A_j is given by:

$$R_j = \sum_i u_{ij} r_{ij}$$

The summation corresponds to all the reactions involved.

Chemical reaction rates are expressed as functions of the concentrations or partial pressures. in the case where two reactants A and B are involved, the kinetic rate is most often expressed as:

$$r = kC_A^m C_B^n$$

Where m and n are respectively the order of the reaction with respect to A, and the order with respect to b. the total order of the reaction is m+n.

For reversible reactions, one has

$$r = k_1 C_A^m C_B^n - k_2 C_P^p C_Q^q$$

And for autocatalytic reactions, this becomes

$$r = kC_A^m C_P^p$$

For complex mechanism reactions formed by a combination of a number of elementary kinetic processes it is sometimes impossible to approximate the chemical reaction rate by one of the three equations above. The following types of rate equations can give better results:

$$r = \frac{kC_A C_B}{1 + K_1 C_A + K_2 C_B}$$

$$r = \frac{kC_A C_B}{[1 + K_1 C_A + K_2 C_B]^2}$$

In the chemical reaction rate equations, effect of temperature is taken into account in the reaction rate constant K which is expressed by:

$$k = k_0 \exp(-E / RT)$$

It can be noted that the chemical reaction rate has been referred to the unit volume of the reaction mixture. However, for heterogeneous systems it can be to base it on another property. The volume of the liquid or the solids contained in the reactor, the mass of the solids or their internal or external surface area between two phases, etc. conversions from one to the other are easily obtained.

Relative degree of conversion, selectivity and yield:

The relative degree of conversion of a reactant A is a measure of the extent of the reaction. it is defined as the function of the amount of the reactant, fed in prior to and during the reaction, which has been converted:

$$\varsigma_A = \frac{N_{A0} - N_A}{N_{A0}}$$

When products other than those desired are formed from the reactants A and B, it is convenient to use the concept of selectivity and yield. The selectivity of a desired product with respect to a key reactant A is the ratio between the amount of P obtained and the amount of A converted. These amounts must take into account the stoichiometry of the reaction.

$$S_p = \frac{N_P - N_{P0}}{N_{A0} - N_A} \frac{v_A}{v_P}$$

The yield of the product P with respect to the key component A is the ratio of the amount of P obtained to the amount of A fed in, corrected for the stoichiometry considerations:

$$\eta_P = \frac{N_P - N_{P0}}{N_{A0}} \frac{v_A}{v_P}$$

By combining equations we obtain

$$\eta_P = s_P \varsigma_A$$

Classification of reactors:

- Homogeneous systems

- Heterogeneous systems

Homogeneous systems are autoclave reactors, tubular reactors. Fixed bed, fluidized bed, mixing bed etc.

Homogeneous systems:

- batch reactor (BR) – under isothermal conditions

- semibatch reactors – under isothermal conditions

- CSTR→ steady state

- ↓ under isothermal conditions

 Unsteady state

- plug flow → steady state

 ↓   - under isothermal conditions

Unsteady state

References

- Pyke, G. H. (1984). "Optimal Foraging Theory: A Critical Review". Annual Review of Ecology and Systematics. 15: 523–575. doi:10.1146/annurev.es.15.110184.002515

- Tomczyk, John; Silberstein, Eugene; Whitman, Bill; Johnson, Bill (2016). Refrigeration and Air Conditioning Technology (8 ed.). Cengage Learning. pp. 518–519. ISBN 9781305856622

- Spiegel, L (2006). "A new method to assess liquid distributor quality". Chemical Engineering and Processing. 45 (11): 1011. doi:10.1016/j.cep.2006.05.003

- Bryan H. Bunch; Alexander Hellemans (2004). The History of Science and Technology. Houghton Mifflin Harcourt. p. 88. ISBN 0-618-22123-9

- Kunesh, John G.; Lahm, Lawrence; Yanagi, Takashi (1987). "Commercial scale experiments that provide insight on packed tower distributors". Industrial & Engineering Chemistry Research. 26 (9): 1845. doi:10.1021/ie00069a021

Permissions

All chapters in this book are published with permission under the Creative Commons Attribution Share Alike License or equivalent. Every chapter published in this book has been scrutinized by our experts. Their significance has been extensively debated. The topics covered herein carry significant information for a comprehensive understanding. They may even be implemented as practical applications or may be referred to as a beginning point for further studies.

We would like to thank the editorial team for lending their expertise to make the book truly unique. They have played a crucial role in the development of this book. Without their invaluable contributions this book wouldn't have been possible. They have made vital efforts to compile up to date information on the varied aspects of this subject to make this book a valuable addition to the collection of many professionals and students.

This book was conceptualized with the vision of imparting up-to-date and integrated information in this field. To ensure the same, a matchless editorial board was set up. Every individual on the board went through rigorous rounds of assessment to prove their worth. After which they invested a large part of their time researching and compiling the most relevant data for our readers.

The editorial board has been involved in producing this book since its inception. They have spent rigorous hours researching and exploring the diverse topics which have resulted in the successful publishing of this book. They have passed on their knowledge of decades through this book. To expedite this challenging task, the publisher supported the team at every step. A small team of assistant editors was also appointed to further simplify the editing procedure and attain best results for the readers.

Apart from the editorial board, the designing team has also invested a significant amount of their time in understanding the subject and creating the most relevant covers. They scrutinized every image to scout for the most suitable representation of the subject and create an appropriate cover for the book.

The publishing team has been an ardent support to the editorial, designing and production team. Their endless efforts to recruit the best for this project, has resulted in the accomplishment of this book. They are a veteran in the field of academics and their pool of knowledge is as vast as their experience in printing. Their expertise and guidance has proved useful at every step. Their uncompromising quality standards have made this book an exceptional effort. Their encouragement from time to time has been an inspiration for everyone.

The publisher and the editorial board hope that this book will prove to be a valuable piece of knowledge for students, practitioners and scholars across the globe.

Index

www.ingramcontent.com/pod-product-compliance
Lightning Source LLC
Chambersburg PA
CBHW062003190326

41458CB00009B/2958